W9-ANP-394
ST. MARY'S CITY, MARYLAND

In *Cognition and Tool Use,* anthropologists Janet and Charles Keller provide an account of human accomplishment based on ethnographic study. Blacksmithing – the transformation of glowing iron into artistic and utilitarian products – is the activity in which they study situated learning. This domain, permeated by visual imagery and physical virtuosity rather than verbal logic, appears antithetical to the usual realms of cognitive study. For this reason, it provides a new entree to human thought and an empirical test for an anthropology of knowledge.

How does a mind in action access a stable, "sedimented" body of knowledge and create something original? What does human tool use say about human thought? What does someone need to know to successfully produce a material artifact and how do they gain this understanding? In addressing these questions, the authors offer an interdisciplinary perspective on the principled creativity of human behavior.

Cognition and tool use

Learning in doing: Social, cognitive, and computational perspectives

GENERAL EDITORS: ROY PEA
JOHN SEELY BROWN

Sociocultural psychology: Theory and practice of doing and knowing
Laura M. W. Martin, Katherine Nelson, and Ethel Tobach (editors)
Sociocultural studies of mind
James V. Wertsch, Pablo del Rio, and Amelia Alvarez (editors)
The computer as medium
Peter Bogh Andersen, Berit Holmqvist, and Jens F. Jensen (editors)
Distributed cognitions: Psychological and educational considerations
Gavriel Salomon (editor)
Understanding practice: Perspectives on activity and context
Seth Chaiklin and Jean Lave (editors)
Street mathematics and school mathematics
Terezinha Nunes, David William Carraher, and Analucia Dias Schliemann
Situated learning: Legitimate peripheral participation
Jean Lave and Etienne Wenger
The construction zone: Working for cognitive change in school
Denis Newman, Peg Griffin, and Michael Cole
Plans and situated actions: The problem of human–machine communication
Lucy A. Suchman
Mind and social activity
Ethel Tobach, Rachel Joffe Falmagne, Mary B. Parlee, Laura Martin, and Aggie Scribner Kapelman (editors)

Cognition and tool use

The blacksmith at work

CHARLES M. KELLER
University of Illinois at
Urbana-Champaign

JANET DIXON KELLER
University of Illinois at
Urbana-Champaign

Published by the Press Syndicate of the University of Cambridge
The Pitt Building, Trumpington Street, Cambridge CB2 1RP
40 West 20th Street, New York, NY 10011-4211, USA
10 Stamford Road, Oakleigh, Melbourne 3166, Australia

First published 1996

Printed in the United States of America

Library of Congress Cataloging-in-Publication Data

Keller, Charles M.

Cognition and tool use : the blacksmith at work / Charles M. Keller, Janet Dixon Keller.

p. cm. – (Learning in doing)

Includes bibliographical references and index.

ISBN 0-521-55239-7 (hardcover)

1. Blacksmithing – Social aspects. 2. Blacksmithing – United States. 3. Cognition and culture – United States. 4. Art and technology – United States. 5. Technology and civilization. I. Keller, Janet Dixon. II. Title. III. Series.
GN436.K45 1996
306.4′6 – dc20 96–3303
CIP

A catalog record for this book is available from the British Library.

ISBN 0-521-55239-7 hardback

For Jean Lave

Contents

Series foreword

This series for Cambridge University Press is becoming widely known as an international forum for studies of situated learning and cognition.

Innovative contributions from anthropology; cognitive, developmental, and cultural psychology; computer science; education, and social theory are providing theory and research that seeks new ways of understanding the social, historical, and contextual nature of the learning, thinking, and practice emerging from human activity. The empirical settings of these research inquiries range from the classroom, to the workplace, to the high-technology office, to learning in the streets and in other communities of practice.

The situated nature of learning and remembering through activity is a central fact. It may appear obvious that human minds develop in social situations, and that they come to appropriate the tools that culture provides to support and extend their sphere of activity and communicative competencies. But cognitive theories of knowledge representation and learning alone have not provided sufficient insight into these relationships.

This series is born of the conviction that new and exciting interdisciplinary syntheses are under way, as scholars and practitioners from diverse fields seek to develop theory and empirical investigations adequate to characterizing the complex relations of social and mental life, and to understanding successful learning wherever it occurs. The series invites contributions that advance our understanding of these seminal issues.

Roy Pea
John Seely Brown

Acknowledgments

Some of the ideas presented here initially took shape in prior publications and presentations, as noted in the Prologue. It was our intention when we began writing this volume simply to string these papers together. However, as we reviewed and reflected on this work and the issues we wished to explore, we decided a collection of articles was inadequate to the task. Nonetheless, we could not have produced this book without the earlier opportunities to publish our ideas. We are grateful for each of these opportunities, for discussions with many colleagues that followed, and for the interactions with students and colleagues in courses where we presented our thoughts as they evolved. Only Chapter 5, "Emergence and Accomplishment in Production," remains closely akin to the articulation of ideas in earlier publications (Keller and Keller 1991a, 1993).

Because of the lengthy duration and interdisciplinary nature of this research, many people have contributed significantly to the development of our ideas. Together we wish to thank all of the smiths who have worked with us over the years and who have taken an interest in what must often have seemed an esoteric endeavor. We wish to thank Michael Brian Schiffer, Donald Norman, Nathan Schlanger, David Herdrich, Leo Hoponoa, Kris Lehman, John Singleton, James Stanlaw, Roy Pea, and two anonymous reviewers for Cambridge University Press for reading the book manuscript or portions thereof at various stages and offering critical commentary. Their insights and queries have allowed us to clarify and elaborate our ideas at many points. We thank the University of Illinois for two sabbatical periods, one for each of us, which we devoted to this project, and for a shoestring grant that facilitated production of the graphics for this book. We also acknowledge the support of the University of Illinois Research Board for Charles Keller's second research period in Santa Fe; the Early American Industries Association, which provided funds for travel to various museums in 1989; and the Scholars Travel Fund for supporting attendance in 1991 at the Conference on Critical Problems and Research Frontiers in the History of Science and the History of Technology. We gratefully acknowledge the photography by Dave Minor and graphics by Chuck Stout that appear in this volume. We appreciate the granting of permission by copyright holders for the reproduction or adaptation of previously published material as specifically referenced throughout this work. Together we thank Jean Lave and Seth Chaiklin, Stephen Levinson and John Gumperz, Dorothy Holland, Lamont Lindstrom, Len Talmy, and Steven Lubar for invitations to participate in conferences and public lectures. To our colleagues who joined us at these events go our thanks for stimulating discussion.

Jean Lave took an interest in our ideas from their earliest publication and encouraged us to continue. Although we sometimes move in directions she probably would not take, we have benefited tremendously from Jean's insights and thoughtful attention over the years. It is in recognition of her dedication to understanding knowledge and practice and in appreciation for her inspiration that we dedicate this book to her.

Janet Keller (JK) wishes to thank Brent Berlin, Paul Kay, and Dan Slobin for nurturing her early interests in both universal and culturally particular dimensions of knowledge. Their guidance created an enthusiasm for research on cognition that persists even as the issues and focuses of investigation have changed. To Jean Lave, Lucy Suchman, Yrgö Engeström, Edwin Hutchins, Marjorie Harness Goodwin, and Charles Goodwin go thanks for constant reminders of the dialectic nature of knowledge and practice as developed in their own research. For hours of collaborative discussion and co-teaching, Kris Lehman earns sincere appreciation. During the summer of 1994, JK spent a month at the First International Cognitive Science Summer Institute. This was a productive time, devoted entirely to the issues developed in this book in a vital intellectual context. She is grateful for the opportunity to teach at this institute, for the questions of participants in her seminar, and for the many formal and informal discussions outside her class. To Len Talmy, who organized the Institute, go her sincere thanks. To Erv Segal, Chuck Frake, Claudia Strauss, Naomi Quinn, Eve and Herb Clark, and Don Norman, JK is grateful for opportunities to talk through ideas and to listen. Many students have read and commented on the manuscript during various stages of production. Attention to the issues they raised has often improved our writing. Finally, to Bill Wood, who is finishing his dissertation on Zapotec weaving as we complete this volume, thanks are due for his inspiration in opening up new dimensions in the study of knowledge and practice.

Charles Keller (CK) expresses his gratitude to his anthropological mentors Robert F. G. Spier and J. Desmond Clark. Although they may not recognize much of themselves in the following pages, Rob taught CK the complexity of nonindustrial technology and Desmond taught him the value of intellectual breadth and diligence. Both demonstrated the necessity of direct experience in this kind of research. Particular thanks for patience and tolerance as well as hospitality in home and shop go to R and B, CK's blacksmithing mentors. None of the research would have been possible without the initial help of JRV, then of the New Mexico State Arts Commission and a friend for nearly 40 years. We have already expressed our gratitude to the many members of the artist-blacksmithing community to whom we have talked, but CK especially wishes to thank J and S, with whom he has shared work space and ideas. CK is grateful to the following people and institutions for their help while examining material in their collections: Tim Talbott and the Early American Museum of the Champaign County Forest Preserve; Frank White and Old Sturbridge Village; Jay Gaynor, Mike Lewis, and Martha Katz-Hyman of the Colonial Williamsburg Foundation; Tom Sanders and the Minnesota Historical Society; and Kathryn Boardman and the Cooperstown Farmers' Museum. For their hospitality and assistance and an unforgettable day as part of the Dominy Family workshop exhibit, thanks go to Charles F. Hummel, Robert Trent, and the Henry Francis du Pont Winterthur Museum.

While teaching courses on archaeology and nonindustrial technology through several decades and in a variety of forms, CK often felt that there was more to be said than he could articulate. Groping for the ideas reflected in these pages was often painful for lecturer and class; his sincere gratitude is offered to those students who endured the process. Special thanks go to Dr. Carol A. B. Link, whose apprenticeship-based research presaged CK's and whose dissertation was important in focusing his research plans.

Finally we both owe our deepest appreciation to Julia Hough, editor in psychology and cognitive science at Cambridge University Press, for her interest as our project was taking shape and for shepherding us and this book through the labyrinthine publication process; to copyeditor W. M. Havighurst for his perceptive and substantive contributions; and to production editor Janis Bolster.

Prologue

This research has grown from a convergence of the authors' lifelong interests in knowledge and technology. The focus on blacksmithing began during a six-month apprenticeship undertaken by Charles Keller with a blacksmith in Santa Fe, New Mexico. CK, as we will refer to him throughout, has continued to practice blacksmithing and to work with other artist-blacksmiths in the United States.

Our collaboration began shortly after CK's return from his apprenticeship in an effort to characterize whatever it is one needs to know to operate in a manner acceptable to other blacksmiths (Goodenough 1957). In this effort we have complemented CK's apprenticeship with both formal and informal interviews of other practicing blacksmiths, observed as they work, and become familiar with their products. We have reviewed available literature on blacksmithing and related activities, much of which is written by blacksmiths themselves. We have also begun research on the role of the blacksmith in early Midwestern America, analyzing the ledgers of smiths working in the 19th and early 20th century and participating in the recovery and analysis of the tools and products from two shops operating during this period.

The two of us have spent long hours in debate and analysis probing the practices and knowledge of both our blacksmithing and our academic colleagues, attempting to create an intellectual framework that would account for the accomplishments of our subjects. Because of the inadequacies of standard anthropological positions in the 1970s we found ourselves disenchanted with cognitively framed explanations of the principled creativity of human behavior. Therefore we have attempted to reach beyond paradigms to integrate subdisciplinary and disciplinary perspectives in the construction of an anthropology of knowledge. Our goal is a unitary volume that uses cognitive anthropology enhanced by activity theory and phenomenology as well as insights from contemporary perspectives on technology and material culture.

The collaborative research and writing that has resulted in this book entails a synthesis of diverse thoughts, biases, and experiences. As authors we complement rather than replicate each other's expertise. CK is both insider and outsider, a "native" through apprenticeship and practice and an anthropologist with a background in the study of stone age technology in sub-Saharan Africa (C. Keller 1966, 1970, 1973; Hansen and Keller 1971; Keller, Hansen, and Alexander 1975). Janet Keller (JK) brings to the work studies of the knowledge associated with Oceanic ritual and cosmology (Keller and Lehman 1991) as well as a study of the symbolism and technology of Oceanic basketry (J. Keller 1988), the latter involving learning to plait from experts.

JK's earlier research, founded in cognitive anthropology of the 1970s, provided an intellectual directive for developing an anthropology of knowledge. Cognitive anthropology of this period, rooted in linguistic expression and universalist semantics, offered a foundation for the comparative study of knowledge systems. Yet while challenged by its goals, we found this framework too static to address questions of the dynamic processes of knowledge construction and use, and too narrow to encompass the disparate substantive components of knowledge associated with production. Interest in these questions led us to branch out and explore the possibilities for developing a broader perspective on knowledge and practice. Early steps toward framing an anthropology of knowledge appear in *Directions in Cognitive Anthropology* (Dougherty 1985), an edited volume compiled by JK to assess research trajectories evidenced in the work of cognitive anthropologists in the early 1980s. Intellectual issues that arose in evaluating this work fed our research on blacksmithing.

The process that has led to the present book was neither simple nor direct. But with Kirin Narayan, who has written so eloquently about the complex relations of anthropologists to their subjects, we hope that in some small way "persevering has brought the reward of greater insight" (1993:678).

Methodology: apprenticeship

Because of CK's involvement in this project as scholar and practitioner, some of the methods used to collect the information on which this book is based may appear unusual at first glance. In fact there is a well-established pedigree in anthropology for the approaches we take. Apprenticeship, perhaps the least used methodology among those employed by us, was applied by Reichard in the 1920s (1974 [1936]) and more recently by others (Link 1975; Coy 1989; Singleton 1996). The reasons for selecting this approach were outlined in a statement written by CK in 1976 as part of an articulation of the goals of his research. At this time CK was interested in exploring the communication of knowledge from expert to novice. In explaining his choice of apprenticeship as a method, he states,

> Both the physical and mental aspects of technological activities become habituated and routinized to the point that information about them is difficult to verbalize. In addition, simple observation of a process often fails to record subtle but crucial aspects of the particular activities. So instead of using the more conventional approach of collecting information from one or more informants about what they think and feel, I am using an apprenticeship approach and recording the way in which I am socialized into this particular activity. (C. Keller 1978)

Apprenticeship is nothing more than the logical extension of the participant-observer method long advocated by ethnographers. The rarity of its application has more to do with the paucity of technological ethnography than it does with any problems inherent in the method. Cognitive anthropologists, in fact, have recognized their field methods as akin to the intensive learning of childhood, a process that shares much with apprenticeships. And anthropologists interested in language have often been concerned to acquire competence in using a language of interest in addition to eliciting linguistic data for analysis.

The apprenticeship portion of this research was begun in February of 1976 in the shop of R, an artist-blacksmith in Santa Fe, New

Mexico. CK continued his work in this shop until July of that year. Since he was interested in the transmission of technological knowledge from an expert to a novice, it seemed necessary that the research should be in a craft with which he was unfamiliar. Although he had done some rather simple jewelry work with silver, iron was a material with which he had no familiarity and so blacksmithing seemed an appropriate activity on which to focus.

CK's initial contact with R had been arranged by a friend who was at the time a Crafts Specialist with the New Mexico State Arts Commission. R was the only one of three blacksmiths contacted who indicated a willingness to take an apprentice into his shop or with whom it was feasible for CK to work. In a brief letter before CK's departure from Illinois for New Mexico, R wrote that the apprenticeship arrangement could last for the period desired, about six months, as long as CK was not a "drag on the shop," and suggested it would be good if CK knew how to arc-weld before he arrived.

The word *apprentice* and its derivatives are used here in a general sense to indicate that there would be no cash involved in the arrangement and that in exchange for CK's contribution to the various projects being carried out in his shop, R would teach him not only what was required to be useful to his efforts but about other blacksmithing techniques as well. From the outset R knew CK was engaged in anthropological research.

CK's nonforging duties in the shop included the following chores: answering the telephone; sweeping the floor; cutting stock to size; cutting to size and drilling holes in brackets used in the installation of porch railings; spray painting and cleaning the spray gun afterward; going to the bed of the Santa Fe River and collecting boulders for a wall R was building; putting tools away after a job was completed; going to the steel supply house for material; assisting in the installation of completed projects (window grills, a gate, several railings); arc-welding decorative elements into the frames of railings or gates; and assembling window grills and cleaning the welds afterward. Forging jobs included making numbers of small scrolls used as the feet for fireplace screens; flattening one end and pointing the other of square bars used in window grills and gates; making scrolls for decorative elements in larger projects; and various other small tasks that utilized his increasing competence at forging.

During the first months, when CK required instruction and supervision while forging, there was no clear distinction between the effort directed to production for the shop and more general instruction. Learning to handle the fire, choose and use appropriate tools, and produce forms of an acceptable quality were basic to CK's research objective as well as required for his contribution to shop activities.

Although in theory all the tools in the shop were available for CK's use, it soon became clear that some tools were personal. Within the first few days he was guided through the manufacture of a poker to use at the forge assigned to him, and directed in the selection and purchase of a forging hammer as well as its modification for use in the shop. As time went on, and depending on the pressure of work in the shop, CK was given some time during the day for his own projects. These were not elaborate and included the production of other tools to be used around the forge, such as a large spoon for carrying coal to and clinker and ash from the fire and a handle for a water sprinkler used to control the fire in the forge. Suggestions were made about techniques to use and possibilities for decoration.

Completed pieces were critiqued, but early comments tended to be monosyllabic and noncommittal; as time passed and CK's skill increased the criticisms became more pointed. During those early weeks CK was observed fairly closely to make sure he was not a danger to himself or anyone else. He received instruction and practice in arc welding (a skill not acquired previously despite instructions to the contrary), proper stance at the anvil, and adjustment of the anvil's height.

R needed some convincing that CK really wanted to be actively engaged in the blacksmithing activities of the shop. More than once he said he had expected that CK would watch what went on and write in a leather-bound notebook. CK was able to convey to him the necessity and genuineness of participation.

The usual daily routine for CK was to drive across town, arrive at the shop close to eight o'clock, and start in on whatever there was to be done. R would arrive sometime later – he lived two doors away –

and give CK any necessary instructions. CK would work through the day, making a mental note of topics, procedures, or comments that seemed important. As soon as he returned home in the evening he made a list of the day's activities and these mentally recorded events; after showering and eating dinner he used the written list as a guide for more detailed notes on the day's occurrences. The advantage of this procedure was that it did not interrupt the natural flow and rhythm of the ongoing activities of the day. The disadvantage was that some detail might have been lost, but this possibility seems minor by comparison with preservation of the cadence and pace of the work.

CK's official working arrangement was with R, but he had considerable contacts with other smiths in the setting of R's shop or as a result of his affiliation with R. Foremost among these was exposure to B, who rented work space from R in a garage-like building adjacent to R's shop. CK discovered later that B was one of the other smiths who had been approached by his friend from the Arts Commission, but B had responded that he had no need for extra assistance. (The third smith approached had expressed mild interest but had been jailed on drug charges before negotiations advanced very far.)

B would often come into R's shop to visit, to take a break, to show something, or to observe CK's progress. He and R would talk about aesthetics, design, workmanship, and similar topics. R usually returned to his house for lunch while both B and CK brought sandwiches, so the two often ate together in R's shop if the weather was cold or outside B's shop if it was sunny. In addition, going to the well, which was the source of drinking water and water for the shop, required passing in front of B's shop. Since B worked near the open double doors of the building, CK often stopped briefly to visit or look at what he was doing if it didn't interrupt him.

CK was invited to B's house for dinner, became acquainted with his family, and eventually spent a fair amount of time with them on weekends and during evenings. B and CK shared common interests in antique tools and firearms and the techniques used to produce them as well as some ethnographic interests and spent many hours in pleasant discussion.

In the last weeks of July 1976, domestic and economic strains on R, combined with CK's own frustration at limits placed by R on CK's blacksmithing activities, led to a deterioration of their relationship. At the same time B was relocating his shop space. CK, who had acquired a basic set of tools, was invited to join B where he had established his own new work space. However, this arrangement did not last very long because of complaints from neighbors about the amount of coal smoke generated and CK's obligation to return to Illinois and teaching.

When CK resumed blacksmithing back in Illinois, it was B whom he consulted for advice when questions or problems arose. B was older than R, had been interested in metalworking longer than R, and was more knowledgeable about historic styles of ironwork, although both were equally proficient at forging processes.

A third smith, M, also rented space from R in the same building as B, but he was there only half a dozen times in the months CK worked at R's and did not contribute in any substantive way to CK's acquisition of forging knowledge. Comments about M by R and B, such as "He's more concerned with varnishing his anvil stump than actually working," were, however, an indication of the values they held and contributed to CK's understanding of the principles that guided their activities.

Beyond R's shop on the far west edge of Santa Fe, another blacksmith ran a school at which six or eight fee-paying students were given practical instruction over a six-week period. On occasion one or more of these students would stop by R's shop to see if there were tools for sale or to show off something they had made. For several weeks R hosted an informal Friday afternoon seminar on various topics such as heat treating or forge welding or multiperson striking. The participants were R, B, CK, perhaps M, a few of the students just mentioned, and sometimes CK's friend from the Arts Commission. Normally these gatherings involved drinking some beer, discussing the topic of the day, and then everyone trying their hand at the procedure of interest.

On one occasion, through the contact at the Arts Commission, R, B, and CK were able to visit the collections area of the International Museum of Folk Art and examine and discuss the ironwork there.

On another occasion a blacksmith from Arizona who had done a number of large commissions, had a national reputation, and had been something of a mentor for both R and B came to town for a visit and spent several days working in R's shop finishing up some projects he had brought along. During this time the normal routine was replaced by stories of work by acknowledged masters and discussions of aesthetics, solutions to unusual problems, sources of coal, and the like.

Finally, sometimes another smith would come to R's shop to visit and to look at equipment or work in progress. And when R and CK were away from the shop together they would sometimes visit the shops of other smiths in the area. Whenever they encountered the work of others they examined and discussed it.

During this period, from February to August, although the apprenticeship arrangement was with R, CK was exposed to nearly a dozen other blacksmiths in varying degrees and settings. Of this group, B made the greatest contribution to CK's blacksmithing knowledge.

CK also spent the following summer of 1977 in Santa Fe working independently, with a meeting every week or so with B to discuss work in progress or ways to approach particular problems. This was a much less intense period of blacksmithing activity than the previous time had been. Contact with blacksmiths was confined to visits with B.

Research and practice beyond apprenticeship: steps toward synthesis

The times just discussed, about eight months in 1976 and about three in 1977, were the only periods when blacksmithing was CK's full-time activity. In later pages we will present statements from a number of blacksmiths that attest to the fascination they find in working with the plasticity of hot iron; CK happily confesses to sharing that fascination. The vital incandescence of the metal being heated in the forge is at once intriguing and challenging, and the aesthetic component of forging became combined with the intellectual facets of the activity for him. Consequently, CK continued blacksmithing on his return from New Mexico: from 1977 to his retirement from the University of Illinois in 1992, he engaged in active blacksmithing more or less constantly, arranged between teaching, writing, and domestic obligations. CK's contact with other smiths expanded considerably during this time but also followed an intermittent pattern.

Yet another source of information on which this book is based is interviews by JK of CK and others. Following CK's return from Santa Fe in the fall of 1976, JK worked with him in an attempt to elicit the organizational structures for tool categories acquired during his experiences in R's shop. What followed were a series of perplexing interactions. CK found the research questions, presented in the style of ethnographic semantics, irrelevant to his practice, as did other smiths informally questioned. JK found the declared irrelevance of the research questions initially disturbing, then ultimately enlightening. It became clear that even were blacksmiths to answer the taxonomic frames posed by ethnographic semantics, the resulting hierarchy would not answer the question, What is it that a blacksmith knows that accounts for his ability to produce acceptable artifacts in iron?

Faced with the realization that a paradigm based on the ethnoscientific classification of the natural world was not adequate to account for the conceptual complexities of goal-oriented behavior, we developed the notion of knowledge as potentially reorganizable according to a task at hand. These ideas were first published in 1982 (Dougherty and Keller) and reprinted in 1985 (Dougherty).

Further thought and discussion between ourselves led to an increasing concern and fascination with the cognitive processes of creative behavior. It was useful to assume that the categories involved in such behavior were organized in shifting, flexible ways. It was also clear that the content and organization of these ideas had to be derived from analysis of dynamic applications during production sequences.

An opportunity arose to directly observe creative production, as well as reflect on the working process, when after about 10 years of using work space in his home in Illinois, CK was offered the chance to share space in a building with two other metalworkers. This became our second most important research site. One of the two other metalworkers, J, had received some training as a blacksmith; after blacksmithing for a couple of years he had become interested in knife making and was primarily engaged in that kind of work when he became acquainted with CK. J had his shop in a first-floor room of an 1870s-vintage building that had been built as a blacksmith shop. Rooms on the second floor were used by S, who had been trained academically in metalwork and jewelry and who had worked during the late 1970s designing and selling custom jewelry. As precious metal prices rose, working capital became an increasing problem and S began using copper and brass to make kitchenware, candelabra, and serving pieces of his own design. In 1986 CK was able to begin working in a first-floor room adjacent to J's, a room that had been the original blacksmith shop when the building was first constructed. The arrangement of three smiths independently practicing in the same building lasted until about 1988, when S moved to larger quarters at another location; in 1989 J moved to another town.

This arrangement was an extremely congenial one. J and S had gone to college together and remained friends after their graduation. J and his wife informed CK of the opening in their building and the whole relocation would not have been possible without their help. J and CK, although working independently, moved back and forth between each other's shops, borrowing tools, contributing an extra pair of hands when helpful, asking and observing what the other was working on, sharing ideas and frustrations. Since S worked upstairs and was engaged in a different kind of metalworking, there was a little less interaction with him, but when entering or leaving the building he usually passed through CK's or J's shop. J and CK usually went upstairs to S's shop for their common lunch break. Ideas were freely exchanged, a question asked of one frequently elicited an additional answer from the other, and mutual stimulation, criticism, and encouragement were the norm.

Information gleaned from observations, initial interviews, conversations, reading, and reflection had been sufficient to frame our questions about the organization of information and concepts during the process of their application, but a more pointed approach was required to clarify the workings of this system. To this end JK, by this time an informed observer-analyst as a result of her years of exposure to and conversations with blacksmiths, interviewed and photographed CK during the production of a particular item. Concern with the issues of knowledge in practice led us to attempt such an interview with CK while he worked. Because interviewing someone during production slows down the work itself, it was not a feasible technique to employ with most practicing smiths. In addition, the interview requires articulation of knowledge often visually or kinesthetically represented, and difficult to verbalize. Verbalization of his practical knowledge is a skill CK acquired as a result of frequent scholarly activities in which explication of his work and that of others was appropriate. The interview lasted a full day and a half and provided a text for analysis developed over the following few years (Keller and Keller 1991a, 1993) that ultimately served as the data base for Chapter 5.

More recent challenges and contexts for research

During the time CK was working in his Illinois shop, JK continued to observe CK and others as they worked, slowly formulating questions that began to elicit information about the underlying knowledge of their craft. As this research progressed and its orientations diversified, JK conducted a number of formal and informal interviews. Some of the material in Chapter 6 on the role of imagery in production is based directly on interview data from work with J.

Another context for research resulted from the gathering of interested persons in situations where forging could actively take place; techniques, equipment, and processes were demonstrated and examples of work exhibited. These gatherings are discussed in Chapter 2; they are mentioned here as another important source of information about the details of the activity as well as the nature of the group engaged in it. Whether involved as a presenter or observer, we found the information that flowed among the participants in these settings important in confirming some of the ideas presented in this book and in revising others. In situations like these commentaries are likely to be made or heard and rules of thumb articulated. From the fall of 1976 CK was a participant in approximately four dozen formal events of one or another kind and JK attended about a dozen events. This participation greatly expanded the number of smiths with whom we have interacted.

The fact that many, if not most, of these gatherings were predominantly active demonstrations is in striking contrast to, for instance, academic conferences where verbal presentations with audiovisual adjuncts are the norm. CK's original apprenticeship demonstrated that much information relevant to blacksmithing was conveyed in modalities other than the verbal, and subsequent experience has confirmed this fact. Attendance at demonstration events has reinforced our notions of the significance of nonverbal reasoning and representation to smithing. This led us to question the linguistic focus of much contemporary cognitive science literature. We became concerned with the diversity of the mental representations held and used by the people we had been working with. The cognitive modalities of these representations and their specificity, interrelations, and recombinations were explored in the interview with J mentioned above. They were first discussed in an academic context in "Imaging in Iron or Thought Is Not Inner Speech," delivered by JK to the 112th International Wenner-Gren Conference on Rethinking Linguistic Relativity in 1991 (Keller and Keller 1996), and in "The Dynamics of Productive Activity," presented by CK to a joint conference of the History of Science Society and the Society for the History of Technology in 1991 (Keller and Keller 1991b).

As CK's experience grew and his familiarity with equipment and ideas developed, the information to be gained from looking at the work of other blacksmiths became more and more precise and useful. He began to undertake research on the knowledge and practices of 18th- and 19th-century American blacksmiths. Over and above looking at individual pieces in museum cases or antique shops, the opportunity to examine specimens in museum collections closely has been extremely important in formulating hypotheses about how particular kinds of tools were made, and then testing these hypotheses at the forge. CK has been fortunate enough to examine collections of 18th- and 19th-century forged iron in the following institutions: Colonial Williamsburg, The Henry Francis du Pont

Winterthur Museum, Farmer's Museum at Cooperstown, Old Sturbridge Village, Mystic Seaport, Historic Fort Snelling, Conner Prairie, Lincoln Log Cabin Historic Site, and the Early American Museum of the Champaign County Forest Preserve.

Examining work produced in earlier years or centuries is extremely informative and stimulating. Using the kinds of knowledge discussed in the chapters that follow, a smith can usually develop ideas about how a particular piece was made on the basis of an examination, but these are really only hypotheses that must be tested and refined by actual attempts at producing a similar piece. Since about 1984 a growing number of the blacksmithing jobs CK has undertaken have involved the replication of period tools and hardware for open air museums and living history sites. This has been a fortunate conjunction of academic and blacksmithing interests, since it has provided an opportunity to test the hypotheses formed during the museum visits mentioned above. The results of some of this work have been summarized by CK in "Invention, Thought and Process: Strategies in Iron Tool Production," published in 1994.

Building on the issues that smiths talked about among themselves and on the information they were willing to share with us, on the problems encountered in our research, on our respective insights into the knowledge and practices of contemporary artist-blacksmithing, and on scholarly critiques of our ideas as they have developed, we established the framework presented in this book. We hope this framework reveals tacit and articulate aspects of the knowledge of blacksmiths and begins to account for their efficacious practice.

Finally, given the research activities we have just described, it is important to point out that this book is not about blacksmithing. We do not intend to explain how to carry out any particular forging procedure or project. Rather, blacksmithing is the empirical base of our research, an investigation of how someone engaged in any of a large number of activities goes about accomplishing his or her goals. The descriptions included in the following pages may at times seem tedious and overly detailed, but they represent the simplest examples we could think of for illustrating the issues relevant to an anthropology of knowledge.

1 Introduction

Our concern in this volume is to understand how people do things. How do individuals with a goal in mind go about accomplishing the end that they see as worthwhile, desirable, or necessary? This question requires examining the workings of the mind in thought and the workings of the body in the physical world.

In the cognitive sciences the investigation of thought is most often conducted by examining speech, a ubiquitous human activity. But humans are also tool users. What might the use of tools, this characteristically human way of doing things, tell us of the workings of the mind? This is the central question addressed in what follows. We will present a theory of cognition derived from our research on productive tool-using behavior among contemporary American artist-blacksmiths; thereby we hope to lay the foundation for an anthropology of knowledge. In our conclusions we extend this theory to a variety of activities to demonstrate its widespread relevance.

Our aim is to establish four major ideas basic to this anthropology of knowledge. The first is that knowledge is purposeful. It is constructed from experience and organized as principles and schemata for attaining one's ends. These ends may range from concrete and material accomplishments to abstract, conceptual, or interactive goals. The second idea is that there is a dialectic between knowledge and practice that gives an emergent quality to accomplishment. We argue that knowledge is both constructed and applied, while practice is simultaneously governed by that knowledge and creative. The process is emergent in that practices have the potential to transcend and reconfigure the traditions and representations on which they are based. This emergent quality of activity we will show is characteristic not only of our domain of study but of human behavior and learning more generally.

Our third idea is that imagery plays a critical role in reasoning; we illustrate this and introduce a preliminary analytical framework for distinguishing among visual representations along both functional and substantive lines. We argue that conceptual thought includes visually and kinesthetically represented ideas that complement and interact with propositional components of knowledge. Finally, by basing our work on the craft of blacksmithing, we establish our fourth idea: that an understanding of material culture is furthered by comprehension of the conceptual processes by which tangible artifacts are created and used. Much day-to-day activity is routine, repetitive, and successful. It represents the consolidation of past experience into patterns and traditions. Yet the potential to meet new conditions exists as a part of the body of conceptual information that enables activity (Holland 1992). Our analysis emphasizes this potential, indicating the conceptual complexity of material production.

Our research is essentially anthropological, rooted in naturally occurring activity and addressing the question of how someone makes something. The cognitive issues of the content and form of knowledge and the dynamic issues of technological production derive from the foundations laid by cognitive anthropology, activity theory, and both ethnographic and archaeological concern with material culture. The issues that become relevant to our account of production, however, also have relevance beyond anthropology in the cognitive science disciplines, where knowledge representation and problem solving are focuses of research.

Knowledge, practice, and structure

We frame this study using the terms *knowledge* and *practice*. By *knowledge* we refer not to a static collection of information, but rather to the disparate and dynamic conceptual entities that individuals use in their various activities. With respect to production, knowledge in our sense does not include all of technoscience, defined as the "principles that underlie a technology's operation" (Schiffer

1992:47), but only that information known to practitioners and applied by them in their work. Our project is an exploration of precisely what *knowledge* is involved in the intentional organization of productive activity. We ask what practitioners know and how they use that information.

By *practice* we refer to the observable behaviors performed in the production of an artifact, the sequences of operations in which individuals engage. The performances we have studied are those of actors involved in making something. Production represents one component of the larger range of "sociocultural practices of a community" (Lave and Wenger 1991:29), which also includes marketing and use. We target producers' behaviors and explore the relationships of these behaviors – referred to as *practice* – with *knowledge.*

For this reason practitioners' behaviors and their insights into the knowledge of their work constitute the crucial data of this study. Our task is to account for the practitioner's knowledge and to explain the place of conceptual thought in production. We do not see conceptual thought as the sole influence on practice but as a governing dimension too often ignored in the study of technology.

Our framework is interdisciplinary and dynamic by design, allowing us to preserve an anthropological foundation but to refine the central problems of our research in light of advances in related disciplines. As a point of departure we take the common premises of contemporary practice theories in anthropology. These agree on the copresence of practice and structure in human behavior, asserting that observed practices emerge from the mental, material, and social structures in which they are situated and, in turn, reproduce or lead to transformations of those same structures (Giddens 1976; Ortner 1989; Sahlins 1985; D'Andrade and Strauss 1992; Sperber 1985; Goodwin and Goodwin 1992; Chaiklin and Lave 1993; Suchman 1987; Bourdieu 1977).

The structures traditionally of most interest to anthropologists are those of the social, cultural, and environmental systems, and these are often treated independently of their representations or manipulations by individual actors. On the other hand, cognitive approaches in anthropology have recognized the significance of conceptual representations and processes for understanding human behavior.

Knowledge constitutes one of the structural underpinnings of practice. For the individual, knowledge is an abstraction from experience integrating the social, cultural, and environmental phenomena that are intentionally brought to bear on behavior. In turn, knowledge is learned in the process of behaving and is organized to facilitate a person's anticipated goals.

By concentrating on the relation of knowledge and practice we hope to avoid the usual pitfalls of structuralist approaches in which behavior is reduced to a determined expression of "preexisting, underlying systems of knowledge" (Hanks 1991:14). We don't want to do away with structure, for cast as dynamic (Hutchins 1980, 1993a; Ortner 1989; Sahlins 1985; Vera and Simon 1993), structures are necessary, yet not sufficient, for an account of everyday behavior. We complement attention to the structural regularities of knowledge and practice with a focus on "proximate" dimensions of production (van der Leeuw 1989), such as intention, choice, reconceptualization, and assessment of unanticipated results.

Creating a focus on knowledge rather than a more typical anthropological focus on systems independent of individual understanding allows us to document precisely the essential role of conceptualization for production. People do not operate within systems of influence that ultimately determine an outcome of their behavior. Rather, people represent, manipulate, and evaluate potentially influential factors in the context of their own goals, purposes, and decisions, to achieve a given end.

Our concern in this book, therefore, will be to address the mutually constitutive relations of knowledge and practice. We will argue that close examination of the ways in which people construct and apply knowledge in making something requires a nondeterministic account of the relations between mind and activity. Knowledge as a resource for production (Suchman 1987:50; Bourdieu 1977; Wagner 1970) governs but does not determine practice; and practices, as they are enacted, may constitute a source of new information and may open prior knowledge to reproduction or transformation with further implications for ensuing practice. This dynamic and constantly emerging set of relations makes possible the characteristic thinking processes of the everyday behavior of human beings and constitutes the foundation for lifelong learning and human productivity.

Our emphasis throughout will be both conceptual, concerned with the representation of information, and situated, concerned with the interface of prior knowledge and a present situation (see recent debates in Greeno 1993). It is the emergent and synergistic character of human behavior that becomes apparent as we proceed. By *emergence* we refer in this work to a person's ability to conceive, act, assess, and reconceive in the process of making something. When new, and sometimes unanticipated, conditions result from actions, knowledge is potentially transformed. Emergence is a characteristic feature of the relation of knowledge and practice.

Building on the foundation of cognitive anthropology (Goodenough 1957; Dougherty 1985) and the potential for cognitive studies of material culture (Renfrew and Zubrow 1994; Schlanger 1990; van der Leeuw 1993; and Trigger 1991), we argue that production is conceptually governed; therefore, attention to individuals' considerations in making something must complement analysis of the social, cultural, and environmental systems wherever possible if our account of production is to be complete.

Practice, activity systems, and implements

Our perspective is derived from an analysis of blacksmithing in contemporary North America. Blacksmithing constitutes an activity system (Leont'ev 1981) incorporating knowledge, readily observable behaviors, and tangible material instruments; it is performed by a group that shares values, information, and goals. Anthropologists often see activities as the empirical manifestation of a society's organization (Schiffer 1992:4). We appreciate this perspective and make a move here to complement this systems theory approach with a cognitive theory exploring the hypothesis that practice is largely the empirical manifestation of the socially derived conceptual organization(s) of the actor(s) involved. We follow our anthropological colleagues in adapting a focus on naturally occurring activity (Singleton 1996; Chaiklin and Lave 1993; Schiffer 1992; Lemonnier 1993; Goodwin and Goodwin 1992). Yet we are optimistic that the account we develop will be applicable to scholars in the more laboratory-oriented disciplines of cognitive science as well (Brown, Collins, and Duguid 1989; Greeno 1993).

The activity system of blacksmithing is attractive for our research for several reasons. Shared foundations for practice among those who identify themselves as artist-blacksmiths provide a general orientation and sense of community. However, within these parameters there is considerable diversity. In part, we wish to account for both the shared and variable dimensions of the knowledge and practice of contemporary American artist-blacksmiths. In addition, the products and practices of blacksmithing are concrete, observable manifestations of and sources for the knowledge we wish to explore, making delineation of an empirical focus clear-cut. The approach taken in this research and the selection of blacksmithing as the activity for study allow us to reduce the complexity of situated cognition and real-world production to a manageable degree by focusing on well-defined, goal-oriented, observable actions. That blacksmithing is well defined does not mean that it is isolated from other activity systems or social realms (Schiffer 1992). It does mean that the focus of our concern can be defined by the practices of production and through these practices can be closely connected to the views of the actors themselves. The test of our proposals regarding knowledge structures governing blacksmithing is whether these can be shown to account for both observed practices and for actors' intuitions about that practice.

In addition, blacksmithing crucially entails the use of tools and equipment in production. Anthropologists have long recognized that humans are distinguished in the animal world by their pervasive, effective, and creative use of tools. Yet few contemporary anthropologists interested in knowledge and practice have focused on actual tool use as an entree to dynamic cognition (Wynn 1991; but see Suchman 1987; Hutchins 1993a; Lave 1988; Lemonnier 1992, 1993; Link 1975; and for a related study with ties to anthropology see Engeström 1993).

In order to take advantage of the centrality of tool use to human nature, we have grounded our research in and built upon the practice theory derived from the insights of L. S. Vygotsky, whose perspective inspired the development of the contemporary scholarly tradition referred to as activity theory. Primarily a developmental psychologist, Vygotsky characterized human intelligence as an essential convergence of instrumental and abstract processes that result from the reactive and transformative relations in which people engage with the world. Human behavior, he argued, not only is a reaction to nature but has that " 'transforming reaction on nature' " which Engels attributed to tools (Vygotsky 1978:61). As a result, to gain an understanding of the uniquely human forms of intelligence, research must be framed in contexts of human activities and tool use.

While, as Vygotsky argued, *tools* can refer to words and other intangibles as appropriately as to material objects, blacksmithing provides an activity in which the tools themselves are easily identifiable and the practices incorporating them easily observed. There can be no question of the significance of tool use for production in blacksmithing. Locating our study at the forge and anvil takes us a long way from the "white rooms" of early cognitive anthropologists, wherein knowledge was elicited untainted by the exigencies of everyday behavior.

Attention to the practices of human production gives a purchase on technological knowledge by aptly characterizing it as conceptual, instrumental, and dynamic. Such an account of technology is more adequate, we will argue, than analyses limited to structuralist foundations, for it allows us to incorporate these foundations in a larger framework of analysis. The practice perspective we are developing entails an integration of data from the study of artifacts, observations of practice, the long-term acquisition of expertise by one of the authors, and informants' accounts of their production. These constitute independent sources of information about the subject of interest, strengthening our inferences where we find independent lines of support.

In developing this perspective, we benefit from the work of anthropologists who have traditionally studied material culture (see Lemonnier 1993; Schiffer 1992) and who have demonstrated the social embeddedness of technology in the process of their research.

These scholars emphasize that activity is the basic human behavioral unit (Schiffer 1992:4). An activity, as Schiffer argues, is defined as

> a patterned interaction between an energy source (e.g. human, machine, animal) and other material elements . . . The people who take part in an activity comprise its social unit . . . [and] activities . . . bind the behavioral and material, the social and biological, and the ideological and technological. (1992:4)

This conception of activity is markedly similar to that of Leont'ev and Vygotsky. However, Schiffer's framework emphasizes features of the social, symbolic, and environmental systems that converge in embedded technologies. He pays less attention to the subjective side of production than do Leont'ev and Vygotsky, both of whom are also interested in the processes of internalization or knowledge construction. Given the usual restriction of archaeological data to material remains, Schiffer's emphasis on the objective properties of technology is appropriate. Our project in this book complements the archaeological interpretation of material culture in terms of systems variables by focusing attention on the conceptual processes of relevance to production.

The integration of information in our research is possible because we have available both material artifacts and the artisans responsible for their production. While we locate our research in a contemporary (dispersed) community, the approach has many of the advantages of ethnoarchaeology and experimental archaeology, allowing us to address the material conditions and artifacts of production in conceptual and cultural terms (Anderson 1993; Arnold 1985; Brumfield 1993; Coles 1973; Ingersoll, Yellen, and MacDonald 1977; Krause 1985; Maryon 1936, 1938, 1941; Peacock 1982; Renfrew and Zubrow 1994).

Principles and schemata

As we approach the convergence of the conceptual side of our dialectic with the practical in naturally occurring adult activity, we have searched outside anthropology for perspectives that would facilitate this integration. Here we borrow from the phenomenology of Alfred Schutz (1967, 1971; Wagner 1970), which is a philosophical account of everyday behavior in terms of associated structures of knowledge and practice. Schutz leads us to expect a stock of knowledge "sedimented" from experience, which, in turn, serves as the basis for future acts. The nature of the processes of sedimentation is left open and to some extent problematized in our own research, but this rooting of knowledge in experience and the orientation of knowledge for experience suggests one of our primary hypotheses: that knowledge is organized for doing rather than abstracted into various formal arrangements on purely logical or typological grounds (see Schlanger 1994; Strauss 1992).

As a consequence of the expectation that knowledge is constructed through processes of sedimentation during experience, we are led to ask what sorts of basic conceptual representations we should anticipate – that is, in what forms might we expect knowledge to be sedimented. Here we turn to recent interdisciplinary research on folk theories of specific domains to enhance Schutz's perspective (Hirschfeld and Gelman 1994). We propose two basic conceptual structures, *principles* and *schemata,* to account for blacksmithing knowledge (see Carey and Spelke 1994; Gopnick and Wellman 1994).

The *principled knowledge* introduced in Chapter 2 is a set of orienting postulates that define the domain of activity for the blacksmith and provide an integral coherence to the diverse projects, techniques, and tools entailed in smithing (see also Gatewood 1985). These principles are abstract relations and methods constructed by smiths over many and diverse instances of making things; they generate a culture of production and its outcomes.

Schemata, by contrast, as referred to commonly in the cognitive literature, are empirical generalizations that represent particular plans, procedures, tools, or artifacts. These are multimodal structures that potentially incorporate visual, kinesthetic, aural, and propositional information. Schemata are represented by actors in terms of the relevant experiences from which they are derived (Gopnick and Wellman 1994). The extensive inventories of schemata a smith might construct from his experiences can be integrated and organized in terms of the more abstract governing principles or immediate goals of a task at hand.

Umbrella plans and constellations

Our practice perspective rooted in activity has been expanded by the incorporation of both conceptual and contextual variables as central objects of study; with this expansion, we can begin to account for the emergent character of production. However, this framework takes us only so far in our investigation: to a stock of knowledge accounting for the possibilities for action, but not yet to an account of the achievement of a specific goal. We will propose that an account of productive activity requires analysis of an overarching *umbrella plan* that defines a goal for production, and further, of a construct that is in essence both mental and material and enables enactment of the plan. This takes us again beyond analyses that would focus exclusively on either mind or matter to see their integration and mutually constitutive contributions to planning and production as perhaps more basic to human activity than either in isolation. We will argue that at the foundation of human production is a unit of ideas, tools, and materials that we call a *constellation.* These units express hypotheses represented in integrated mental and material form for the achievement of a step in production. Constellations enable actions with reproductive and transformative potential for the constellation itself, for the material artifact at issue, and for the umbrella plan and stock of knowledge from which the constellation is derived.

The openness of knowledge, plans, and constellations to revisions based on the material results of enactment recall the structure of the conjuncture proposed by Sahlins (1985) to account for historical change. Sahlins indicates that "by the 'structure of the conjuncture' I mean the practical realization of the cultural categories in a specific historical context, as expressed in the interested action of the historic agents, including the microsociology of their interaction" (1985:xiv). Similarly, we find a task at hand provides an occasion, a situation, for the practical realization of conceptual notions. A task at hand presents a problem. The smith conceptualizes his goal and plans his strategy for accomplishment. Then in each step of the activity the smith hypothesizes a means to the end of the current step, realizing this hypothesis in the constellation of tools and materials assembled for work and testing it in the procedural enactment that follows. Here ideas are at risk in practice (Sahlins 1985:ix) and practices are the source of new ideas as well as reproductions of established notions. This recognition of the tension between ideas and material practices is essential to an account of emergent synchronic events.

Previous integrations of knowledge and practice

We have focused herein on the relations between knowing and doing in production because these relations have consistently arisen as problematic and criterial issues in our research over the past 15 years (Dougherty and Keller 1982; Dougherty 1985; Keller and Keller 1993). Attention to either knowledge or practice to the exclusion of the other has rendered prior accounts of each component inadequate. We are joined in this endeavor by a number of colleagues who have likewise been challenged to integrate knowledge and practice in the search for increasingly adequate accounts of human behavior (Harper 1987; Lave 1988; Barker 1968; Chaiklin and Lave 1993; Dougherty 1985; Jordan 1993; Hutchins 1993a; D'Andrade and Strauss 1992; Gladwin 1970; Holland and Eisenhart 1990; Holland and Quinn 1987; Ortner 1989; Sahlins 1985; Singleton 1996; Suchman 1987; Gorman and Carlson 1990). These scholars come from diverse disciplinary backgrounds and locate their research in very different empirical domains, but all share a recognition that a complex dialectic is central to an account of human activity. Some of these scholars emphasize one or the other side of the dialectic. Others are more interested in the nature of the relations themselves. And differences are evident in the precise construction of the terms of the dialectic. However, all of the researchers noted above are committed to understanding human behavior as a complex, multidimensional, and relational process.

It is useful to situate the perspective we have developed here with respect to this body of similar research. The ecological psychology of Barker (1968) is one of the earlier attempts we note to characterize a dialectic foundation for human behavior. Reacting against an exclusive focus on the disembodied mind characteristic of so much psychological research, Barker accounted for activity in terms of social and environmental settings. He recognized two mutually constitutive positions by which to orient his project. First, an individual acting "is a component of the supra-individual unit [behavior setting], and secondly, he is an individual whose life-space [individual psychology] is partly formed within the constraints imposed by the very entity of which he is a part" (1968:17). It is this dialectic between life-space and behavior settings that recalls our project in this book. Yet Barker tended to proceed by delineating the structural properties of behavior settings to the exclusion of the dynamic processes occurring in these settings and their effects on the individual's life-space. In removing the focus of psychology from the disembodied mind, Barker demonstrates the significance of contextual structures for human behavior. Yet while he acknowledges the importance of reintegrating these with the processes of the mind itself, his research never quite takes this step.

More recently Lave (1988) has demonstrated the effect of settings for behavior and arenas of activity within those settings on individual psychology. Her research has inspired the development of a scholarly tradition emphasizing the situated construction of activity. However, Lave too, while a leader in acknowledging the significance of a dialectic between persons acting and contexts of practice (both settings and arenas), still builds her own account almost exclusively in contextual rather than conceptual terms (Hutchins 1993b; Lave 1988; Lave and Wenger 1991; Suchman 1987).

Other scholars, developing approaches out of ethnomethodology and the sociology of frame analysis (Goffman 1974), have focused on interaction and collaborative accomplishment (Goodwin and Goodwin 1992). Like Lave and Barker, they "view actors as not simply embedded within context, but actively involved in the process of building context" (Goodwin and Goodwin 1992:97). They emphasize that interaction involves a mutually constitutive process by which the evaluation of activity becomes part of that very activity (1992:83). This introduction of evaluation into the dialectic provides a mechanism by which the role of conceptual processing becomes critical to understanding persons acting and thus broadens the dialectic of behavior and context set up in the work of Lave and Barker. Intentionality and representation become crucial to such analyses of conversation.

The recognition that conceptual processes are central to human behavior suggests our own focus on the conceptual processing at issue in production. Yet it complicates the analysis of human behavior, for we ultimately have three sets of phenomena to consider: the conceptual, the behavioral, and the contextual. It is at this point that we have chosen to emphasize the dialectic between knowledge and practice for a given actor. We recognize that there is a larger and more encompassing set of dynamic relations among activity, cognition, and the social world. In fact, as we complete the final editing of our volume, Ed Hutchins's new book, *Cognition in the Wild* (1995), has just appeared. In this work he explores distributed cognitive systems that integrate individual cognition, artifacts, technological practices, and communicative interactions in the accomplishment of the tasks of navigating large naval vessels. Our work accounting for how someone makes something, by contrast, explores in depth the conceptual and productive dimensions of a more individually reliant activity system. Yet as the analysis we develop touches on the social and environmental contexts of production, we hope to provide another vantage point for understanding the larger picture as well.

Our approach must be further distinguished from the work of scholars who target interaction for investigation. Dialogue lends itself to analysis as a collaborative and intersubjective accomplishment. This framework, while often suggestive in the study of human–material activity, cannot be adopted uncritically, for only the human in this latter relationship is a volitional subject (cf. Schlanger 1994:148). Human–material activity is more properly understood as a dialectic rather than a dialogic process. Mental and material dimensions are synthesized in production. It is this dialectic that we develop in our focus on tool use.

A number of contemporary anthropologists interested in cognition and in the question of the relationship between culture and knowledge have developed approaches similar to our own. Many ask, among other questions, a version of one originally articulated by Goodenough (1957): What is it one needs to know to behave appropriately in a particular situation? (Holland and Eisenhart 1990; D'Andrade and Strauss 1992; Holland and Quinn 1987; Schlanger 1994; Quinn 1985; Hutchins 1993a). These scholars focus on the knowledge underlying human activities – knowledge simultaneously acquired and applied in activities such as factory work, stone tool production, getting a college education, maintaining a marriage, or navigating a vessel. They address issues of the substance, form, and distribution of mental and material representations, situating these at a dynamic interface with experience. Much of this research is ongoing, and like our own, still developing. However, while we find ourselves closely allied with these scholars, there has often been an emphasis on the representational side in their work, an emphasis we hope to complement here by integrating the representational and behavioral components of production.

Ultimately our focus on the dialectic of knowledge and practice will require expansion to allow this process to be situated in its contexts of influence. However, the dialectic of knowledge and practice that characterizes production cannot be explained alone by the systems or contexts that influence it. These must be understood, integrated, and manipulated by an agent. It is, therefore, toward an account of the agent behaving that we turn. Here we must work initially from the actor outward to avoid contextual determinism, recognizing that it is equally crucial for others to work from the structures of context inward. No comprehensive account of human behavior can take one of these perspectives to the exclusion of the other. It is toward an integration of complementary positions on what we will argue is a unitary phenomenon that we are moving.

Significance of the knowledge–practice dialectic

A number of implications for anthropology and related disciplines derive from the framework we have developed. First, the anthropological concern with ideal versus real behavior appears in a new light when structure and performance are understood as mutually constitutive. Analyses of ideal accounts address structures independent of action. Holding all else equal, an ideal expresses what one should do. Yet if structure and behavior emerge in a mutually constitutive process, ideals or norms for behavior can be understood as applicable in particular circumstances. The circumstances are as essential a component of the ideal as are the governing principles. Alternatives will emerge as circumstances change. Both ideal and alternative strategies are equally real, each enacted as deemed appropriate in context.

Secondly, while this book is about competent performance, it entails the potential for a reconsideration of learning. Both reproductions and transformations in individual knowledge and practice over time constitute learning. This is not learning limited to childhood or to episodes of novice status in adulthood (see also Hill and Plath 1995). The learning at issue here is an ongoing characteristic of the human condition (Chaiklin and Lave 1993; Lave and Wenger 1991). One learns how to accomplish a present goal in an emergent process by integrating a current situation and knowledge of past experiences. The learning at issue is a continual building of substantive principles and empirical generalizations with incremental, reproductive, and transformative possibilities. Performance in this view is as integral to what someone knows as that knowledge is to their performance.

Further, our approach to knowledge and practice assumes a cognitive theory that takes as its central problem establishing the universal architecture of the mind. This architecture constrains ways of knowing and applying knowledge, and shapes the constructive learning process (Jackendoff 1989; Hirschfeld and Gelman 1994). However, our goal is to focus on the substance of the knowledge for a particular domain of experience and on the diversity of cognitive modalities that contribute to the representation and implementation of this substance in practice. Ultimately such substantive analyses should provide evidence for the basic architecture of the mind.

Sperber and Wilson have argued that cognitive processes are designed to let you know as much about the environment as possible (1986:47). We agree and will argue here that what one knows about the environment is oriented to what one does in that environment; once again this will require a dual focus on both mental representations and practical actions. In this we are making a contribution to the study of cognition by reinforcing the ties between mind, behavior, and matter. We are casting cognition as inherently goal-oriented knowing for doing, with a view toward clarifying the connections between mental models and what people do.

As we pursue the dialectic relationship between knowledge and practice in emergent production, we are led to ask what forms are taken by an actor's knowledge representations and what the nature of the material influences on them might be. Here we highlight the significant role of imagery in visual as well as other modalities. A part of our task in this book will be to clarify the role imagery plays in production and to explore the interactions among cognitive modes of representation with particular emphasis on language and vision (Jackendoff 1992). Informants themselves constantly reminded us that such imagery is critical to stable, learned structures of knowledge, to new representations a blacksmith constructs in context, and to successful production. Images of general applicability in blacksmithing combined with task-specific needs of production may be used by the blacksmith to create imaged plans, hypotheses, and schemata for reasoning about particulars of design or procedure. Such images tested in practice are subject to revision on the basis of the results of enactment as directly perceived. Task-specific imagery subsequently feeds back into general knowledge of the domain constructed in visual or other imaged form.

We find that the knowledge responsible for effective blacksmithing broadly encompasses diverse forms of mental representation. Conceptualization is perhaps an appropriate overarching rubric for the thinking that enjoins activity and it encompasses visualization and kinesthetic imagery as well as language and emotion. The capacities of the mind are differentially designed for the representation of distinct forms of information (see Jackendoff 1989; Pinker and Bloom 1990; Hanks 1991). A full understanding of how someone makes something entails an account of each contributing mode of information processing and the integration of these processing modes in conceptualization (Keller and Keller 1993). While this task is more comprehensive than we can accomplish here, we initiate consideration of the role of mental imagery in production because of its critical importance to successful blacksmithing and, by extension, to other thoughtful activities (Bloch 1985; Gatewood 1985; Hutchins 1983; Shore 1991; Strauss 1992).

Not only are we moving toward a unified theory of cognition and behavior, but we are motivated in this project to account for the subjective unity of work, "the marriage of the hand and the mind in solving practical problems" (Harper 1987:118). It is only through relating the knowledge and practices of production that we are able to begin to account for this unity of experience.

We also find that an approach that essentially integrates knowledge and practice avoids the oversimplification of "form follows from function" so common to studies of technology and design (see critical reviews in Gorman and Carlson 1990; Pfaffenberger 1992; Pye 1978). Knowledge of both forms and functions influences planning and production. Our reliance on conceptualization as crucial to production demonstrates the inadequacies of positing a unidirectional causal relationship.

Finally, developing a conceptual foundation for productive activity challenges linear analyses of production sequences. Recourse, for example, to a chain of operations observed during production cannot provide an adequate account of an activity. To address the questions raised in this book, we must begin by discovering operations as understood by practitioners. Production of the sort we examine is governed by and carried through in the intentions and actions of an individual – through plans, constellations, and reassessment. Practitioners remain open in their work to new conceptualizations, techniques, and strategies (Lemonnier 1992, 1993; Renfrew and Zubrow 1994), and material results of action are open to novel developments as a result of an agent's reconsideration of a goal or act (Hardin 1984). The process of production is not linear but dialectic, entailing constant movement from conceptual to material structures and back again.

The chapters that follow

Because this book takes a relational perspective within an even broader perspective that deals with a person-acting-in-the-world (Lave 1988), the point at which the study begins is somewhat arbitrary. To discuss any one aspect of the system one must establish a focus that relegates related phenomena momentarily to the periphery. However, to progress in understanding how someone makes something we need to begin somewhere, and we have chosen to follow the lead provided by the smiths themselves, that is, to access the stock of knowledge through the organization of the shop and then to move on to their forging practices. To conclude, we will address interrelations among elements of production initially treated in isolation. We will also develop the general implications of this volume in the final chapter by reference to other activities, such as invention (Gorman and Carlson 1990; C. Cooper 1991b; Uselding 1974), craft (Lechtman 1984; Hardin 1984; Link 1975; Singleton 1996), and industry (Vincenti 1984; Lubar 1987), in which we find performance is similarly a dialectic of knowledge and practice.

Throughout this book we refer to blacksmiths with the male pronouns *he, him,* and *his.* Not all artist-blacksmiths are male, nor do we believe the framework we propose is in any way unique to men. However, these pronouns are empirically accurate, for our research has relied exclusively on artist-blacksmiths who happened to be men. This reflects the predominance of men in the activity of forging.

Following this Introduction, Chapter 2 summarizes the social and cultural dimensions of contemporary artist-blacksmithing. We address issues of knowledge and community here and contextualize the activities targeted for study. A number of additional issues that might have arisen in a fuller analysis of blacksmithing are not addressed in this chapter. In particular, the issues of communication between clients and smiths, while relevant to a more complete ethnography, is precluded here. The diversity of the clientele for whom blacksmiths work and the nature of their requests is sufficiently great that we would be significantly diverted from our purposes by trying to account for the blacksmiths' communication task in identifying, satisfying, and creating the objects of customers' requests. We restrict Chapter 2 to a discussion of those dimensions of artist-blacksmithing that follow the initial decision to make something.

Chapter 3 addresses the stock of knowledge revealed through the organization of a shop and the inventory of its tools and equipment. This chapter deals with issues of the sedimentation of knowledge from experience and the representation of that knowledge in terms of anticipated future experiences.

Chapter 4 introduces the umbrella plan and constellations as the enabling foundation for actually doing something. Chapter 5 further develops the concept of the umbrella plan and demonstrates the emergent character of production. This includes a discussion of the implications of emergence for continual expansion and revision of structures of knowledge and practice. Here we wrestle with the interface of situated practice and the representations of knowledge discussed in the preceding chapters.

Chapter 6 addresses the role of mental imagery in the blacksmith's general stock of knowledge and in the dynamic processes of planning and production. In the concluding chapter we take a step toward articulating a framework for an anthropology of knowledge transcending disembodied minds and rooted in the study of human activity and artifacts.

2 A profile of artist-blacksmithing

Blacksmith is defined by *Webster's Third New International Dictionary* (1981) as "a smith who works in iron with a forge." This definition applies generally to farriers, industrial house smiths, fabricators, and artist-blacksmiths. Taken together, these groups constitute diverse communities of blacksmiths active today, each distinguished by its products and procedures.

Spectators at events where a blacksmith is demonstrating the craft often make comments like, "Oh look, it's a blacksmith, he's making a horseshoe," or "There's a blacksmith, he's probably getting ready to shoe a horse." Statements like these, based on historical recollections or on movie and other mass media depictions, reflect a contemporary stereotype of the blacksmith as one who shoes horses. Historically there is some validity to the association of the task and the person, although an examination of historic documents indicates that shoeing horses, technically farrier work, was never the main occupation of blacksmiths, let alone their only occupation.

Today horseshoers or farriers are set apart from other blacksmiths for a number of reasons. First, their primary activity, shoeing horses, requires a knowledge of horse anatomy and behavior not shared by blacksmiths whose activities do not include shoeing. Secondly, many farriers move their work location to the racetrack or boarding stable, where their customers' horses are kept. This mobility results in equipment that is smaller and lighter and different in its design, nature, and energy requirements from that of other ironworkers. Thirdly, many farriers use shoes purchased from a manufacturer and heated and shaped only to the degree necessary for the final fitting of the shoe to the hoof. Consequently, the farrier changes the

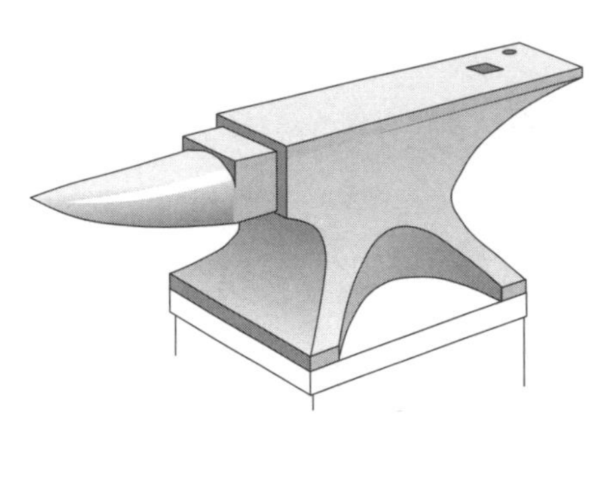

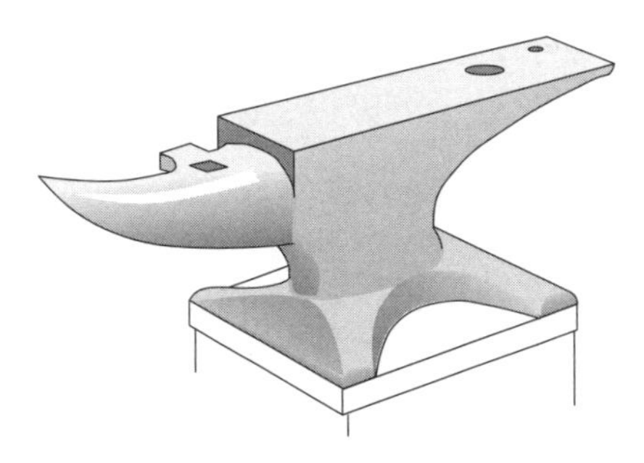

Figure 2.1. Anvils typically used by blacksmiths: English-pattern anvil (top) and farrier's anvil.

shape of the work piece less than is often the case with other blacksmiths. This also has implications for the farrier's equipment, since the temperature required for most of his work is lower than that of other smiths. Finally, the anvils used by farriers possess a number of specialized features that make them readily distinguishable from the more common English-pattern blacksmith's anvil (see Figure 2.1).

Another distinction within the realm of blacksmiths is that between industrial or house smiths and those who work in small shops or for themselves. Industrial blacksmiths may work for a forging company that uses closed dies and large drop hammers to mass-produce items such as hammerheads, wrenches, axes, shovels, gears, and any of a myriad of implements and components described or labeled as being "drop-forged." In companies of this kind the blacksmith also makes, modifies, and repairs tongs used to hold the heated work piece under the drop hammer and he repairs broken or bent pieces of equipment. Other blacksmiths who work in industrial settings may actually use drop hammers themselves to forge products, as for shipyards or lumber mills.

A third contrast is that between fabricators and other smiths. Some contemporary decorative "ironwork" involves no forging whatsoever and consists of arc- or gas-welding together cast elements or cold-bending thin stock and welding it into railings, stanchions, or grills. Individuals who do this kind of work are usually referred to as "fabricators."

The blacksmiths on whom we focus in this book identify themselves as artist-blacksmiths. In the founding statement for the Artist-Blacksmiths' Association of North America (1973), Gerakaris, one of the founding members, declares, "We understand that a blacksmith is one who shapes and forges iron with hammer and anvil. The artist-blacksmith does this so as to unite the functional with the aesthetic, realizing that the two are inseparable" (1993). He goes on to announce, "Our task is great and so is our joy" (1993). This emphasis on creativity, effective procedures, working products, and the joy of forging iron is characteristic of artist-blacksmiths.

These individuals typically work by themselves or with a few others in small shops where the techniques of earlier smiths are combined with contemporary tasks and equipment. Artist-blacksmiths represent the present-day practitioners of blacksmithing as it was frequently done in the late 19th and early 20th centuries. Then, one or two people worked in a shop carrying out the repair and specialty production required by their community. These artisans were resources for the community and were often called upon to invent material solutions to problems arising in everyday activities. Today's small shops also tend to involve one or a few smiths and to produce a range of products, although contemporary smiths are typically more specialized in their work than their predecessors. Among contemporary specialties are historic reproduction, architectural work, and edged tool production. Some shops, for example, focus on re-creating the tools and hardware of earlier periods for museums, historic sites, collectors, and decorators. Others work primarily with architects, builders, and property owners to produce custom architectural ironwork such as gates, railings, balustrades, and window grills. These products may range in style from traditional work derived primarily from that of the 18th and 19th centuries in Europe to forms that are at the forefront of contemporary aesthetics. Another subset of those smiths working in small shops are cutlers who make knives and other edged tools for outdoorsman, collectors, and woodworkers. Examples of the work of contemporary artist-blacksmiths appear in Figures 2.2–2.5.

For thousands of years iron has been transformed by blacksmiths into utilitarian and artistic objects. The noted metallurgist Cyril Stanley Smith said of iron, "There is no metal in which the nature of the material interacts with the craftsman's hand and eye more profoundly and to better effect" (Smith 1966:29). The dynamism of this interaction between smith and material creates a "love for the work" almost universally professed by artist-blacksmith practitioners. Talking about his own work, artist-blacksmith John Creed says, "Using an elemental material and shaping it rapidly, allowing things to grow . . . You just take some metal, you put it in the fire, which is an elemental thing, take a hammer and you work it there – it is so *honest*" (Andreae 1993). The basic processes of forging become consuming. Other artist-blacksmiths echo these feelings. Daryl Meier says, "I just got hooked on this [forging] and I'm going to stay with it until I pass out of the picture" (Reichelt 1988:76). Brent Kington says, "I found my way to another material [iron] and saw in it a wonderful potential, something I just couldn't resist investigating" (Reichelt 1988:102). From Darold Rinedollar: "I really miss it if I don't come in here [to the shop]. If I go somewhere I feel antsy. I just want to get back and go to work" (Reichelt 1988:58).

At the root of this love of the craft are a number of intertwined dimensions. One is perhaps the way in which one project stimulates another. "One thing opens up the door to somewhere else. You can't work fast enough to accomplish it . . . There's just not enough time

Figure 2.2. Forged hardware for a toy box. Charles Keller, 1984.

Figure 2.3. Door hardware. Brian Anderson, 1976. Reprinted by permission of the University Museum, Southern Illinois University, Carbondale, from *Iron, Solid Wrought/USA* (1976), p. 40. © 1976 University Museum and Art Galleries, Southern Illinois University, Carbondale.

Figure 2.4. Portal gates. Albert Paley, 1974. National Museum of American Art. Reproduced by permission of the National Museum of American Art, Smithsonian Institution, Commissioned for the Renwick Gallery.

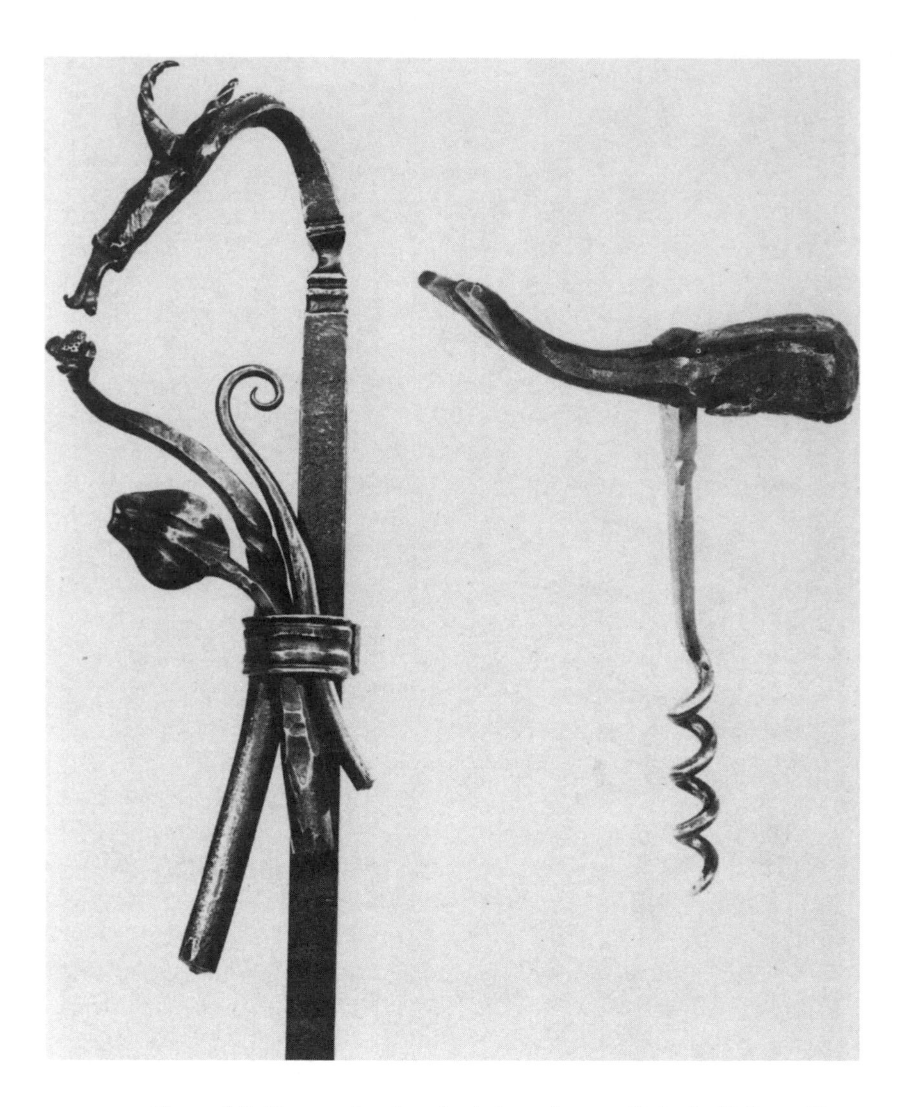

Figure 2.5. Dragon sketch and whale corkscrew. Rolando DeLeon. Reprinted by permission of the University Museum, Southern Illinois University, Carbondale, from *Iron, Solid Wrought/USA* (1976), p. 27. © 1976 University Museum and Art Galleries, Southern Illinois University, Carbondale.

in your lifetime to do everything that you have ideas for" (Darold Rinedollar in Reichelt 1988:58). The work also calls for a unity of mind, hand, and material that intensely engages the smith in his work. A brief comment from Francis Whitaker (1986: xiii) is illustrative:

> If the eye cannot see it
> The hand cannot make it.
> If the tongs will not hold it
> The hammer cannot hit it.
> If you cannot remember it
> Forget it.

Novelty in production too is a source of satisfaction for artist-blacksmiths. Combining the importance of novelty with an emphasis on the unity referred to above, Richardson, writing in the 1890s, says the blacksmith cannot make the most of his opportunities "unless he has studied, and finds new themes in every heat, spark, or scale. If he can create beautiful forms in his mind, and with his hands shape the metal to those forms, then he can see poetry in his work" (1978 [1889–1891]:3:163).

For many smiths, another pleasure derived from forging is the continuity felt with artists of the past through the influence of past technologies and products on contemporary work. Christopher Ray argues that "there is a sense of reassurance knowing that one has a direct linkage with a past, that this knowledge which is a gift, has been passed on, person to person, an individualized kind of history spoken with the hand" (Meilach 1977:iv).

Finally, underlying all these statements is a fascination with the material itself.

> It was the plasticity of iron and steel that initially fascinated me about them, and attracted me to them. This inherent characteristic served as catalyst to the experimentations that gave me an understanding of these materials resulting in a design theory founded in paradox. The dichotomous nature of these materials present opposite and seemingly contradictory states at the same time and place; movement and stagnation, rigidity and plasticity. The tool imprints; incisions, tears, twists and burns, record the evolutionary nature of process and form development. Movement basic to the organic, of which we are all a part, made visible in the steel becomes a foil to human gesture, resulting in empathy and anticipation through this visual dialogue. (Paley 1993)

The nature of iron and blacksmithing

Iron is one of the most common elements in the earth's crust. It reacts readily with two other elements, oxygen and carbon. Iron oxides are a frequently encountered iron ore and are converted to the metal by driving off the oxygen during the smelting process. In the inverse process, metallic iron will oxidize rapidly when heated, and this influences the conditions under which it is forged. The carbon content of metallic iron is also a significant factor in forge work. At elevated temperatures, carbon is dissolved in the crystalline structure of iron: depending on the amount of carbon present, its arrangement within the crystal, and the speed with which the metal is cooled, materials with different properties result.

Metallic iron is usually found in one of three forms. Wrought iron contains less than 0.03% carbon by weight; steel contains 0.45%–2.0% carbon; and cast iron, 2.0%–4.5% (Sanders 1993:12). Wrought iron and steel have the property of being plastic at temperatures between 1,250° F (675° C) and 2,700° F (1,475° C); they are the raw material of the blacksmith. In addition, steel can be hardened and tempered to produce tools with effective cutting edges, or with the resiliency required for springs. Cast iron is too brittle to be forged, passing from a solid to a liquid state without becoming plastic. Cast iron must be melted and poured into a mold, a procedure carried out in foundries rather than in blacksmiths' shops. However, blacksmiths may occasionally repair items made of cast iron.

Smelted iron appeared as early as c. 4000 B.C. in the Middle East (Wertime and Muhly 1980:69). The ore was converted to metal in small furnaces that produced a spongy slag-filled mass called a bloom. The iron in the bloom was concentrated, and impurities were removed by repeated heating and hammering. Forging in this way elongated the bloom into a bar and the iron and remaining slag were arranged into the solid fibrous unit now known as wrought iron. Initially rare and highly valued, iron was first used for jewelry. As its advantages for tools and weapons became apparent and its production more common, iron was used over an ever-widening area. By about 800 B.C. it was known throughout the circum-Mediterranean area and Europe (Wertime and Muhly 1980).

Many of the forging techniques developed in the six thousand years since the production of iron became common either compensate for or take advantage of the properties of wrought iron. Tools excavated from Iron Age sites in Europe and dating to well before the Christian era would not look out of place in the shop of a 20th-century blacksmith. As Ewbank (1849:231) says, "The tongs, anvil and hammer of Vulcan . . . have come down to our times."

Wrought iron was the blacksmiths' most commonly used raw material until the 20th century (Wertime and Muhly 1980). It contains 1%–3% slag, or iron silicate, by weight; as a result of the manufacturing processes employed in its production this slag is distributed through the iron in the form of long fibers. It is this fibrous nature that many forging techniques either utilize or correct for. For instance, an eye in the end of a rod was often made by bending stock in a circle and welding the end to the body of the rod, leaving the fibers unbroken and retaining their original strength. Wrought iron is also referred to as being "red short," meaning brittle and therefore unworkable, at temperatures that produce a red incandescence, although it forges readily at higher temperatures that produce an orange, yellow, or white glow (see Figure 2.6).

Around the end of the 19th century, production of wrought iron declined in the United States and eventually ceased altogether. It was replaced by "mild steel," which has a carbon content of approximately 0.4%. Mild steel lacks the fibrous nature of wrought iron and also lacks the hardenability of carbon steel. But it is more readily machined than wrought iron and can also be forged over a wider range of temperatures. Mild steel is the material usually manipulated by the blacksmith of the 20th century and it is commonly referred to as "iron," although that is incorrect in the metallurgical sense. However, because this is the blacksmiths' usage, in what follows the term *iron* will refer to mild steel, while *steel* will refer to the carbon steel used for springs and edged tools.

To take advantage of the plastic potential of iron the material must be heated to a temperature between approximately 1,200° and

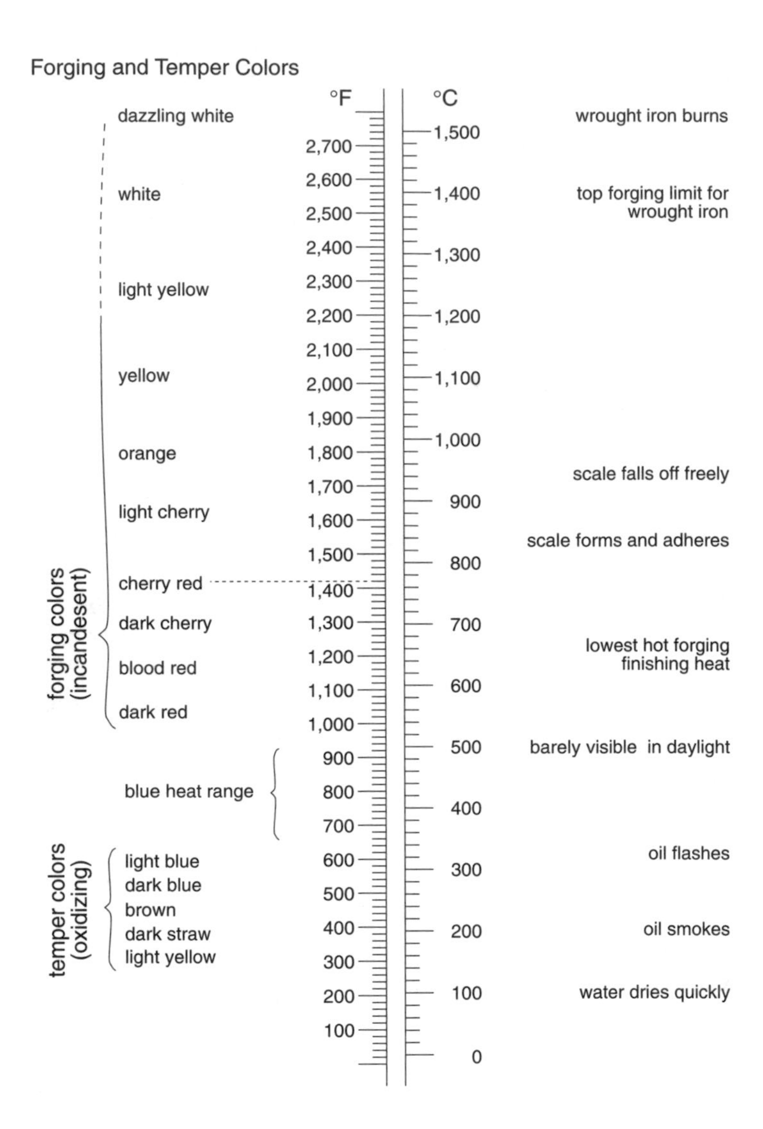

Figure 2.6. Forging and temper colors. Redrawn with the author's permission from Jack Andrews, *Edge of the Anvil* (1977), p. 118 (Emmaus, PA: Rodale Press). © 1977 Jack Andrews.

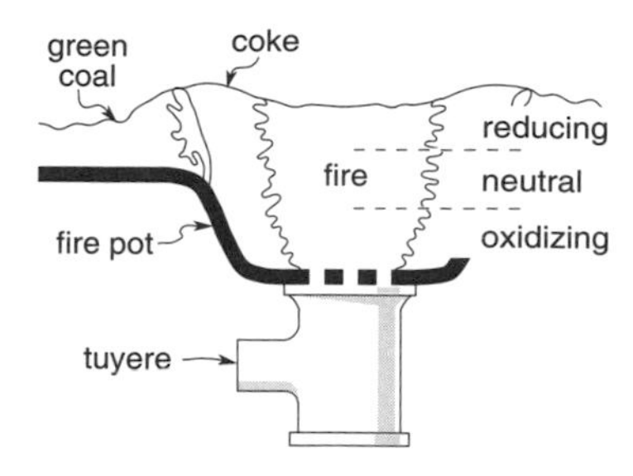

Figure 2.7. Artist-blacksmith's forge showing fuel and the fire. (The tuyere is a nozzle through which an air blast is delivered to the forge.) Redrawn with the author's permission from Jack Andrews, *Edge of the Anvil* (1977), p. 43 (Emmaus, PA: Rodale Press). © 1977 Jack Andrews.

2,500° F. This requires a forge such as that illustrated in Figure 2.7. The forge is made of brick or cast iron and supports a cast iron basin, the fire pot, that has an opening in its bottom through which air is admitted to the fire and from which the by-products of the fire, ash and clinker, can be removed. The fuel of the forge is either charcoal or bituminous coal, which is fanned to the requisite heat by air from a bellows or blower. If bituminous coal is used, it must be capable of being readily converted to coke, the relatively pure carbon produced when the fire drives off volatile gases and impurities (Sanders 1993:16). The fire pot must be large enough to allow the oxygen introduced to be consumed before it comes in contact with the iron or steel work piece. Iron heated in an oxygen-rich atmosphere oxidizes rapidly, which may impart an undesirable surface texture to the work and can interfere with welding two pieces together.

The fire is gaseous, constantly changing, and ephemeral, and so does not conform to the usual conception of a "tool." However, it is the heat of the fire that creates the plasticity required for iron to be forged, and the forge and the fire it contains are basic tools of the craft. It is possible to rapidly alter the size, shape, temperature, and atmosphere of the fire to suit the immediate needs of the work at hand by adding or removing fuel and by changing the amount of air fed into the forge. This mutability is part of the essence of blacksmithing activities. The pervasive objective of smiths is creative transformation. The mutability of the fire at the heart of forging is reflected in the blacksmiths' approach generally. Both the tangible tools and the conventional procedures of the smith are conceived as adaptable to work in progress. Forging itself is the transformation of a piece of stock to a finished product. The notion that nearly everything relevant to forging is subject to change means that the blacksmith works with the potential for reconfiguration in his technology, strategies, and production simultaneously. This mutability, extensive as it is, is constrained by the commitment of most smiths to the idea that their tools should not be used in a way that injures them beyond the effects of normal wear and tear.

Another characteristic of blacksmithing is the intense concentration required to create and maintain the proper configuration of the fire and other tools and equipment for an ongoing project. The condition of the fire and the condition of the iron being heated must be constantly monitored by the smith to make sure that there is sufficient fuel to heat the piece in the desired atmosphere, that the fuel is in the proper location in the fire pot to receive the air blast, that an accumulation of ash or clinker is not obstructing the air flow, and that the work is being heated at the correct rate. Large pieces must be heated slowly and evenly so that the exterior and interior are at approximately the same temperature when the piece is removed for forging. Partially forged pieces that have both thick and thin cross-sections must be heated carefully to avoid overheating and ruining the smaller portions of the work. This need for constant monitoring demands the smith's attention throughout a productive sequence even while other considerations such as proper procedures or the location of required tools may also be at issue.

As the metal is heated, its temperature is judged by its color (see Figure 2.6). At about 900° F (500° C) a slight incandescence can be seen if the iron is held in a shadow. This is sometimes called black heat or planishing heat, since the surface of the work can be smoothed with light hammer blows at this temperature. A "low red" color appears at about 1,200° F (650° C): this is the lowest forging heat, that is, the temperature at which the cross-section of the work piece can be modified significantly by forging. At the so-called cherry red heat, about 1,500° F (810° C), iron and steel become nonmagnetic. At about 1,600° F (875° C) iron oxide forms on the surface of the work and scales off when the iron is exposed to an oxygen-rich atmosphere either in the fire or in the air. The color spectrum continues on through orange (1,800° F [980° C]) to light yellow (2,300° F [1,260° C]), where two pieces of iron can be welded together, to white (2,600° F [1,420° C]), which represents the forging limits of mild steel. Above this temperature a molecular change takes place and the material crumbles if struck with a hammer. The application of color as an indicator of malleability will be discussed more fully in relation to other systems of visual knowledge in Chapter 6.

When metal has been heated to the required temperature, it is removed from the fire and "forged." Forging refers to the techniques of transforming material through thinning, thickening, shortening, lengthening, narrowing, spreading, cutting, piercing, splitting, twisting, or bending. While some of these techniques would be possible if the metal were cold, they could not be done to the same degree, nor with the same final appearance, and some of them could not be done at all. Iron worked cold lacks the surface texture, changes in cross-section, and flowing lines that are the desired hallmarks of hot work and a source of satisfaction for blacksmiths and their customers.

The basic implements for forging, in addition to the fire, are the anvil, hammer, and vise. The anvil is the tool on which the metal is placed or that supports other tools on or against which the metal is placed. A hammer is used to drive the hot metal into, against, or around the anvil or its ancillary tools, or to drive cutting or piercing tools into the hot iron. A vise is used to hold work for bending, twisting, chiseling, filing, or straightening, and also to hold other tools used in these processes. There are a myriad of supplementary tools employed in forging, such as tongs, chisels, fullers, hold-downs, punches, drifts, and files; these can all be, and frequently are, produced using the basic equipment of forge, anvil, vise, and hammer.

The nature of forging iron is such that the metal must be alternately heated, then forged, then reheated. As soon as the iron is removed from the fire to begin a forging operation, it begins to cool and thus to lose plasticity; it must be reheated, forged again, and so on until the desired end is obtained. Consequently, the operations of the blacksmith are inherently rhythmic and episodic. One smith has referred to the "dance" performed as he moves from forge to anvil and back again (Tom Sanders, personal communication, 1993). This is strikingly different from working wood, clay, or fibers, where the operations can go on more or less continuously and where it is the artisan who controls the pace of the work. The speed with which iron cools depends on the temperature to which it was heated and the size and shape of the material. A ½-inch-square bar heated to approximately 2,500° F remains hot enough to forge for about 60 seconds. A bar ¼ inch thick and 1 inch wide, which is the same mass as the ½-inch-square bar, will cool more rapidly, staying hot enough to forge for only about 35 seconds.

Several things follow from these physical realities of blacksmithing. First, there is a time period that smiths often refer to as a "heat," which represents the amount of forging that can be done from the time the metal is taken out of the fire until it has to be returned to the fire for reheating. Second, the episodic nature of the activity, alternately heating and forging, divides the operation into a series of natural and discrete units. Third, this alternation of active and monitoring activities provides an opportunity for evaluation of previous acts and planning for subsequent acts. Finally, the fact that the temperature and plasticity of the metal are declining from the instant the work comes out of the forge means that there is an urgency and immediacy to the activity rarely found in other crafts. It is important for the smith to know what he intends to do and to have the required tools at hand before the metal is taken out of the forge.

Community and communication among contemporary artist-blacksmiths

The frequently solitary nature of the artist-blacksmith work of today, the fact that blacksmiths are often widely separated geographically, and the draw their work has for them has resulted in the creation of a number of channels through which information about techniques, projects, and solutions to problems can be exchanged. It is this nexus of communication media that allows a sense of community to be established and maintained. The situation is not unlike that of an academic who finds herself the only representative of her specialty on a given campus. She is bound to her fellow specialists by common research interests, methods, and values, and communicates with them through various media, but she meets them face-to-face only occasionally.

Although there were more blacksmiths in the past than there are now, direct communication among them was usually restricted to the training of an apprentice by a more accomplished person or the information imparted to an assistant by the owner of a shop. Consequently, there is no large body of literature from the past to which contemporary smiths can refer for information about the way earlier blacksmiths went about their work.

Exceptions include the descriptions of equipment and techniques in Moxon's *Mechanick Exercises* (1975 [1703]) and articles in the various encyclopedias published in the early 19th century (Knight 1877), which may have been read by some smiths. It seems likely, however, that none of these significantly conveyed information to the practitioners of the craft, in part because of the limited availability of the publications, which were intended for "gentlemen" rather than artisans, and in part because of the incompleteness of the information they contained. From the late 19th century until about 1920 there was at least one trade journal, *The Blacksmith and Wheelwright,* which was probably more useful to the practicing smith. Also during this period a number of books were published, to some degree "how to" books, that contained information considered helpful for a farmer who needed to make or repair tools or equipment (Friese 1921; Drew 1915; Hasluck 1912). These typically emphasized small tasks like making a hook or repairing a log chain, but some contained descriptions of more ambitious projects such as forging a connecting rod for a steam engine. Somewhat later, a few volumes were published that were intended to be used as texts in shop classes; these are much more thorough in their treatment. One of these at least enjoys considerable popularity among beginning blacksmiths today, even though it is available only in a photocopied form (Schwartzkopf 1916).

Communication and the possibilities for a shared culture among blacksmiths have changed, however, since 1973, when the Artist-Blacksmiths' Association of North America was founded (Gerakaris 1993). This organization publishes a quarterly magazine, *The Anvil's Ring,* in which contemporary artisans contribute articles on techniques, equipment, and solutions to problems they have encountered, as well as numerous illustrations of contemporary work. Recently another publication, *The Blacksmith's Journal,* has appeared; it consists almost entirely of drawings with little text, and it emphasizes tools, techniques, and projects. Local and regional organizations often have newsletters that contain pertinent information, and these are additional resources for contemporary smiths. Also appearing since the 1970s have been a number of books that either reprint information from earlier publications (Richardson 1978), discuss the history of blacksmithing (Lasansky 1980), or present discussions of basic techniques (Weygers 1974; Bealer 1976; Andrews 1977; Meilach 1977). This body of literature presents information about how present-day smiths solve problems of holding the work piece or laying out the work or building or making equipment or tools not available commercially, or any of the other myriad difficulties that may arise in the course of a project. These publications also illustrate the work of contemporary smiths, sometimes with accompanying descriptions of how a project was carried out. This literature disseminates information within the blacksmithing community without requiring individuals to travel long distances for face-to-face contact. The information communicated provides a common basis for the construction of standards for practices and products by which members of the artist-blacksmithing community evaluate one another's work. However, all the media of communication mentioned above have the inherent limitations of any attempt to describe a complex and dynamic process with the written word or to illustrate a three-dimensional object on a printed page.

When a smith has the opportunity to examine a piece of work produced during an earlier time or by a contemporary, answering the question, How did he do that? furnishes information about approaches and techniques. While it may not be possible to determine specific details about the tools used in production, the general nature and sequence of the operations employed as well as standards of workmanship in another time or place can be "read" by one smith from something made by another.

By far the most fruitful medium of information exchange for today's artist-blacksmiths is face-to-face communication in the form of workshops, demonstrations, and meetings. These range from a biennial week-long meeting sponsored by ABANA, attended by smiths and featuring multiple and simultaneous demonstrations by selected smiths, to local workshops and "hammer-ins" of one or two days' duration. The latter may be organized by a local group or held at the whim of an individual. It is really only in these situations that the nuances and complexities of techniques and equipment can be appreciated and the quality of workmanship in a displayed object understood.

This conjunction of communication modes provides the context for learning and the construction of identity. An individual's stock of knowledge relevant to blacksmithing grows continuously and is transformed through practices that are understood and evaluated in the context of a framework derived from the ideas and products of others. The processes by which knowledge is sedimented in practice entail both social standards and personal experience. No one is self-taught, although much learning may take place in isolation.

Principles and compromises shared by artist-blacksmiths

Artist-blacksmiths, while not a localized group, do constitute a community by virtue of their identity, their intercommunication, their practices, and their sharing of a worldview based on a set of constructed principles for work. These principles are articulated explicitly in blacksmiths' conversations about their work and are expressed implicitly in their products, performances, and literature.

The competent smith draws on a set of orienting principles for the production and evaluation of work. Historically rooted and currently reconceptualized in the contexts of modern blacksmithing, these fundamentals shape the way artist-blacksmiths approach the unification of the "functional with the aesthetic." It is, at least in part, the resourceful reliance on this worldview that creates the artist-blacksmith's characteristic satisfaction with his work.

The principles outlined below constitute a set of cultural premises that define the domain of artist-blacksmithing (Keller and Lehman 1991; Murphy and Medin 1985). Such principles are significant conceptual structures for activity (Jordan 1993; Gladwin 1970; Hutchins 1983; Hill and Plath 1995) and constitute a foundation for the acquisition of more specific schematic knowledge. The principles represent the ideals of the community members and direct a smith's initial approach to a project. Compromises occur when the exigencies of time or economics push the smith to violate some principle to achieve a practical end.

It is important to note here the differences between principles and the rules often adduced when discussing the production of tangible artifacts (Hill 1994; Schiffer 1992: 46). As will become clear below, we are referring to general premises, ideas constructed by blacksmiths. In the case of artist-blacksmithing the principles constitute axioms of a folk theory of their craft (Keller and Lehman 1991). Artist-blacksmithing is the freehand transformation of hot iron for the creation of objects of beauty and mechanical efficacy. Rules, on the other hand, as typically described, are specific means–ends relations, as in recipes for action or specific strategies for problem solving.

Principle 1: Transformation

It is common to refer to different technological activities by the way material is treated. For instance, chipping stone, carving wood, and machining metal are regarded as subtractive processes since some of the raw material is removed to achieve a particular shape and size of the finished product. By contrast, other activities are regarded as additive in that pieces of raw material are combined in one fashion or another to produce the desired end. Thus welding and carpentry are seen as additive, as are weaving and making pottery by coiling. While some additive techniques, such as welding, and some subtractive techniques, such as filing, are used by blacksmiths, the characteristic approach of the artist-blacksmith is to change the shape of the raw material by transforming it so that, for all practical purposes, no material is lost. Even when welding is used, it is frequently applied only to produce a sufficient mass to allow subsequent transformation of the metal into a desired shape. The artist-blacksmith regards his raw material and conceives of his task in terms of the mass required for efficient production of a desired shape. This can be contrasted with a carpenter, for instance, who may shorten or combine pieces of lumber but who does not normally reshape the standard components with which he works.

This primary transformative approach to production is based on the plasticity of iron, which results from its being heated. Forging hot iron alters the cold and static qualities of the standard-dimensioned bars, rods, and sheets that come from the mill. It replaces those qualities with a depth, a surface texture, an appearance of warmth and flow in the lines and shapes created. It is as though the transformative process gives life to the ironwork. Just as the fire was described as flexible and adaptable and the hot iron as plastic, the smith's frame of mind itself has a potential for elasticity and mutability.

Principle 2: Think hot

A requirement of the transformative approach to metalworking is that blacksmiths shape iron while it is hot, and it is the resultant plasticity of the material that the blacksmith exploits. This contrasts with other crafts such as silver- and goldsmithing, where the material is sometimes shaped by forging, but at the ambient temperature. Blacksmiths sometimes forge thin sheet iron when it is cold, but typically only for special aesthetic purposes. They also do some kinds of ornamental ironwork cold, as by bending thin straps of metal around jigs to form scrolls or other decorative elements used in railings, grills, or stanchions. But "cold work" lacks the texture and changes in cross-section that artist-blacksmiths think of as characteristic of their products.

"Thinking hot" is the defining principle of artist-blacksmith forging, governing the construction of procedures and implementation of techniques in practice. It is the general means to the blacksmiths' transformative ends. A hole in a bar will be hot-punched or slit and drifted to size rather than drilled. Stock will be heated and cut on the hardie (a sharp-edged tool fixed in the anvil over which hot iron can be severed) rather than sawed or cut with an oxyacetylene torch (a 20th-century device that burns oxygen and acetylene gas at a temperature sufficient to melt iron). Grooves will be formed with fullers rather than with files or a lathe. Primary transformations in shape are achieved with hammer and anvil. For example, a fireplace poker can be made by forging a handle on one end of a square bar and forging the characteristic point and prong on the other end. The straight section of uniform dimension in the center can be decorated with a twist or forged to a taper or transformed from square to round. By the time the poker is finished it has been forged in its entirety. No portion is left cold and unchanged. The artist-blacksmith strives to produce ironwork in which the original stock has been transformed in line and cross-section.

The importance of this principle was emphasized during the early days of CK's apprenticeship. In the struggle to produce a desired shape, he would often fail to notice that the material had cooled past the point of incandescence. "It's too cold, Charlie" was a phrase often repeated by one or another of his mentors. This was frequently followed by a reference to negative qualities attributed to those who "pound cold iron."

The high value placed on this principle is illustrated by a comment made by one spectator to another at a blacksmithing demonstration. Two farriers, at their respective anvils, were each attempting to complete a horseshoe by forging one from a bar in a single heat. Neither had completed the task by the time the shoes ceased glowing, but they continued hammering. An arriving spectator asked what was going on, and was told in a disgusted tone by a spectator smith, "It's an exhibition of pounding cold iron."

Principle 3: Working freehand

Blacksmiths usually take a freehand approach to the transformation of their material from stock to finished product. That is, a desired outcome is produced as completely as possible by hand and eye, using a minimum of jigs or templates. Changes in cross-section or line are produced with the hammer and anvil or bending fork and are checked for fairness against sketches and mental images. This preference for working by hand reflects the fascination, challenge, and gratification associated with the intensive process of hand forging. The need for constant attention, the union of hand, eye, and material felt in every blow, the potential for new ideas to grow as work proceeds, all result from working by hand. It is in this handworking that the characteristic delight of the blacksmith arises.

Part of the gratification and often the frustration of working freehand is that it demands skill. By skill we mean a combination of sensitivity to visual and tactile input from the ongoing work, muscular control that allows the efficient production of a desired end, and sophistication of procedures implemented to carry out the work. A skilled performance requires that an actor bring his knowledge of smithing to bear in monitoring and evaluating changes brought about by his actions. Deciding if those changes conform to images of how the work should look at any given stage is part of skilled monitoring. The shape of the piece after each hammer blow as well as the feel of the tool on the material are critical in these assessments. Recognizing problems, diagnosing the cause, and applying appropriate corrective procedures are part of the knowledge-based strategic component necessary to operate skillfully. The manual ability to move the work through its necessary phases as quickly as possible with a minimum of physical errors is important as well.

Working freehand requires skill. Smiths rarely if ever use this word, but their appreciation for skillful practice is evidenced in numerous ways. Sometimes one smith refers to another's ability to accomplish a certain amount of work in "one heat" as an indication of skill. Novices may be told that they should be able to do a certain part of a job in "one heat," implying that they are expected to work with speed and dexterity.

The goals artist-blacksmiths set for themselves entail skilled practice. For example, the ability to precisely and repeatedly select the angle at which a chosen portion of the hammer face strikes the work requires skill; this is referred to as "hammer control," which a smith may acquire, lose, or regain. One smith has said that he wants his work to look like "it grew that way" (Hap Harthen, personal communication, 1988), meaning that the surface of the work should not reflect the mechanisms of its transformation. The achievement of flowing form and uniform surface requires considerable hammer control.

Many of the tools used by blacksmiths call for substantial dexterity on the part of the smith. David Pye (1968:9) has described such tools as lying toward the "risk" end of a spectrum ranging from "tools of risk" to "tools of certainty." While neither end of the scale is intended to suggest an absolute, the terms refer to the degree of mind, eye, and muscle coordination required to use a tool effectively. At the risky end of the spectrum are those tools demanding greater skill and mastery for effective use. A dentist's drill is an example of one of Pye's "tools of risk," since nothing about the form of the implement itself contributes to its guidance or constrains its movement. Tools of this kind, intimately associated with the freehand approach, have the disadvantage of requiring considerable effort to utilize effectively, but they have the advantage of being relatively unspecialized, that is, a single tool can be used to produce a wide range of results.

A preference for tools and techniques that allow a smith to interact as directly as possible with his material is maintained in spite of the labor-intensive situation it creates. Many blacksmithing techniques for forging a piece of iron will produce an approximation of a desired shape but will also result in some undesirable feature as well. For instance, the technique of "drawing out" a piece of stock refers to reducing its thickness and increasing its length. The technique requires only the most basic equipment: the heat of the forge, a hammer, and an anvil. But when a smith thins and lengthens the stock its width also increases to a slight degree. Assuming that the desired shape requires a constant width, the smith must then forge the width back to its original dimension as a secondary or derived part of the technique (Schiffer 1992). If a smith were going to make hundreds or thousands of pieces drawn out to a single set of dimensions, it would be worthwhile economically to make a swage, or a form into which the plastic metal would be forced, and thus avoid the increase in width that occurs in drawing out by hand and the consequent secondary forging step. However, artist-blacksmiths rarely if ever choose working situations in which extensive, exact repetitive production is the goal.

When repetitive elements such as scrolls or leaves are required, instead of relying on a device to constrain the transformations of the metal as it is forged, the artist-blacksmith prefers to use the more demanding and risky hand-controlled tools, even though this entails compensating for any undesired secondary features that may appear. As a result, if one compares hand-forged work with similar cast or stamped-out products, the hand-forged result usually contains irregularities and asymmetries not present in mass-produced items. It is these textural features that blacksmiths recognize as the characteristic qualities of "warmth," "life," or "flow." One smith, watching a novice attempt to produce two identical scrolls freehand, said, "They don't have to *be* identical, they just have to *read* as identical" (Brian Anderson, personal communication, 1976). In fact, reading as identical is perhaps the more desirable effect, for it is in the creation of works that read as integrated, where symmetry of line, balance, and repetition are achieved properties, constructed in forging, that the blacksmith finds satisfaction. Factory productions with more exact specifications involve little of the synthesis of mind, hand, eye, and material so engaging for the smith and leave the resultant product "cold," "dead," "untouched," and "sterile" by comparison.

Each smith must constantly balance the realities of his economic situation with the value of working freehand. Using tools or techniques of greater "certainty" in one part of a project may allow more time for freehand work in another part. This area of discretion is apparent in the remarks of two proficient and commercially successful smiths. One said, "If you need more than six of something, make a jig" (Jack Brubaker, demonstration, 1985). The other, referring to decorative bolt heads forged into a variety of flowerlike shapes, said, "I must have made a thousand of these and no two of them were the same" (Francis Whitaker, demonstration, 1981). Each smith, guided by the principle of thinking hot and the value placed on transforming the shape of the metal, will determine the extent of freehand work in accordance with relevant constraints of skill and economics for each project undertaken.

Discussion

Artist-blacksmiths pursue the novel, the imaginative, and the original in their work. Yet this pursuit is constrained by a shared set of principles and by a commitment to respect the integrity of their tools and the properties of the materials with which they work. Artist-blacksmithing unites the conceptual and material in a process through which products, techniques, and ideas continually grow and develop. Given the principles of transformation, thinking hot, and working freehand, the smith negotiates a path toward the production of an original work.

Compromise is an essential component of the work process. As mentioned, principles represent ideals that may be sacrificed to the exigencies of time, skill, or economics. For example, two pieces to be joined by a forge weld may be "tacked" together with the gas or arc welder to hold them in position for subsequent heating and welding. In a 19th-century shop an assistant would have been available for the procedure, but economic factors today preclude this luxury. "Tacking" violates the freehand principle but facilitates the transformative and freehand process of forge welding and enables production within reasonable limits of time and money.

The process of adapting principles to the constraints of a task or situation at hand and compromising to maximize both the principle-based direction of activity and the satisfaction of practical requirements is typical of the blacksmith's reflections and anticipations regarding his work.

Richardson's comments quoted earlier spoke of finding "new themes in every heat, spark or scale"; John Creed spoke of "allowing things to grow"; Brent Kington of iron's "wonderful potential"; Darold Rinedollar of "one project stimulating another." This constant growth and change forms a context for creativity that has continuity over time and space as a result of the common folk knowledge of artist-blacksmithing. This knowledge is constructed by smiths themselves and is rooted in the founding principles of thinking hot and transforming iron by the skilled application of risky and unspecialized tools. In striving to unite the functional and the aesthetic in their work, the community of artist-blacksmiths finds common ground. We noted in the introduction that knowledge is purposeful. Here we find a set of defining principles that establish the criteria for a theory of artist-blacksmithing. This knowledge in its most basic form is oriented to accomplishment.

Creativity is characteristic of the North American artist-blacksmithing community. Similar patterns can be seen in the work of smiths from societies as diverse as the Merovingians of southern Germany (Koch 1977), the people of Benin (Zirngibl 1983), and the Lozi (Sieber 1972) and Mande (McNaughton 1988) of sub-Saharan Africa. We suspect the processes of principle-governed production and compromise oriented toward constrained originality would equally well characterize the knowledge and practices of the blacksmiths of these groups. Exploitation of opportunities for creativity seems to be a common feature of blacksmithing communities. As we will argue later, generalization to other activities is equally viable, accounting for a wide range of tool-using behaviors.

3 Work space and the stock of knowledge

We begin an examination of accomplishment in forging by focusing on the stock of knowledge reflected in the organization of a blacksmith's shop. Concentrating on the dialectic between knowing and doing (Harper 1987; Lave 1988; Chaiklin and Lave 1993; Dougherty 1985; Mauss 1927 in Schlanger 1994), we ask how the knowledge of smithing is reflected in the acquisition of tools and the establishment of locations for those tools in the workplace. As we have noted, blacksmithing is goal-oriented (Habermas 1971; Schutz 1971; Miller, Galanter, and Pribram 1960; Schank and Abelson 1977). Our analysis rests on the premise that the smith's knowledge is applied in organizing a shop to facilitate the goal orientation of work and the accomplishment of desired ends.

As the smith acquires a tool inventory and organizes that inventory in a shop, he builds the environment in which he works. This process constitutes a practice of the smith in the same way that forging a piece of iron does. We ask how these organizational practices reproduce or transform prior knowledge structures.

We will argue that one can infer basic knowledge of smithing from the smith's collection of tools and their arrangement in a work space. We will focus on two dimensions of the stock of knowledge: first, the evidence for principled knowledge of forging provided by a shop's organization; and second, the procedural knowledge that can be recovered from the tool inventory itself.

Our basic unit of analysis in this chapter is the shop as an expression of a blacksmith's knowledge of his craft. Given the solitude in which blacksmiths typically work, the organization of a shop often reflects the knowledge and expectations of a single individual.

But as we pointed out in Chapter 2, solitude is not to be mistaken for isolation. An individual artist-blacksmith is a member of a widely dispersed social group, yet his decisions, practices, and influences derive from the common orienting principles of that group. Social acts can be performed by a single individual as well as by an interacting set of people. In the case of artist-blacksmiths, adherence to the common principles we have described is more important in maintaining group membership than the physical presence of other persons (Spradley and McCurdy 1972; Lave and Wenger 1991).

The stock of knowledge as defined by Schutz (1971:146) is "The sedimentation of previous experiencing acts together with their generalizations, formalizations and idealizations. It is at hand, actually or potentially, recollected or retained, and as such the ground of all our protentions and anticipations." The organization of a shop, far from being arbitrary, embodies the potential for productive activity and as such is a material realization of the blacksmith's expectations and anticipations regarding future work. We explore the organization of a shop, including inventories of tools and their usual locations, in an effort to reconstruct this underlying knowledge. The inventory and location of tools and equipment on which we base our analysis are general-purpose manifestations of the stock of knowledge. What the blacksmith knows about these tools and their arrangements is dynamic, continually subject to growth and transformation in practice. The stock of knowledge as we reconstruct it represents knowledge at a moment in time, but it is the open quality and potential for reorganization of the information emphasized in subsequent chapters that enables selective retrieval and creative production.

In approaching this project we recognize that discussion of a person's stock of knowledge tackles a subject matter of immense proportions – the sedimentation of all experience and the ongoing growth and transformation of this sediment. It would be as impossible to provide a complete analysis of an individual's stock of knowledge as it is to do a complete ethnography (Bateson 1958). Our aim herein, therefore, is to provide an analysis of aspects of the stock of knowledge relevant to blacksmithing, as these can be inferred from analyses of real-world manifestations of that knowledge. We assume the tool inventory and its arrangement in the shop constitute an artifact of the practices involved in organizing the shop and we use these data to illustrate the substance of the stock of knowledge.

On the basis of two particular shops, we articulate the governing rationale by asking what the practicing smith must know to account for the constructed orders that can be observed. From this we argue more generally for the kind of knowledge a blacksmith must have in order to account for the inventory and arrangements of tools and materials in his shop. Less formal observations of other shops confirm the organization represented here. Our assumption is that the knowledge we reconstruct from the tool inventories and locational arrangements is, for the practicing smith, simultaneously derivative and generative of those very inventories and arrangements.

The importance of the contents and physical layout of a shop as a reflection of underlying knowledge is confirmed in Harper's discussions of Willie, a contemporary northeastern American mechanic-engineer. In his discussions Harper argues, "The shop . . . operates by a logic that, while not obvious at first glance, ensures that Willie remains in control of his time and energy" (Harper 1987:4). The logic referred to here is Willie's stock of knowledge, on the basis of which he has constructed an order for his working environment that facilitates his productive activity. Willie himself, when speaking of the accumulated tools and materials in and around his shop, suggests, " 'To everyone else it looks like junk; to me it looks like stuff I can use' " (Harper 1987:54). Here again it is Willie's stock of knowledge, a sedimentation of previous experience, that leads him to collect a particular array of tools and materials in anticipation of future projects.

The artist-blacksmith, likewise, does not go to a store or a catalog and order a set of standard tools. Beyond the basic implements of anvil, forge, vise, and hammer, most tools and equipment will be acquired singly as needed or encountered, or will be made by the smith himself as the occasion arises.

We were directed to the significance of the inventory and arrangement of tools as a manifestation of underlying knowledge early in our research. One of CK's duties as an apprentice was to clean up after a job was completed. This mainly meant putting away the tools in their proper places to facilitate subsequent retrieval for another job. Learning about tools by locating them in a shop we found is a common element of learning blacksmithing. For example, in discussing the apprenticeship of Charleston blacksmith Philip Simmons, Vlach points to the importance of learning the placement of tools for acquiring basic knowledge. Vlach quotes Simmons himself commenting on the significance of tool location at the beginning of his involvement with a master smith: "I had a lot of responsibility . . . Puttin' things away, puttin' tools in the proper place. I was handy when needed. I had to hand him different tools and hold different things. Old Man say, 'Where's my punch?' I get it" (Vlach 1981:20).

This is an example of learning from the periphery to the center of an activity, a process that has been well observed in the work of others (Singleton 1996; Lave and Wenger 1991; Hutchins 1993a) and that points to tools and their organization as an entree to the knowledge of blacksmithing.

Organization of a shop: implications for the stock of knowledge

We will examine the organization of two blacksmithing shops in detail. The first is the shop in which CK was trained as an apprentice in Santa Fe, New Mexico. The shop structure was built by the owner-smith, who also acquired and arranged the tool inventory. The work done in this shop is primarily ornamental and architectural. The second shop is CK's current work space in central Illinois. It was built in the 1870s by a blacksmith whose business included horseshoeing, production and repair of agricultural and household items, and wagon and buggy repair. Jobs undertaken there today are primarily replication of colonial-period tools and hardware.

The Santa Fe shop

Reflecting on his own apprenticeship, CK reconstructed the tools and their proper places in the Santa Fe shop in which he was

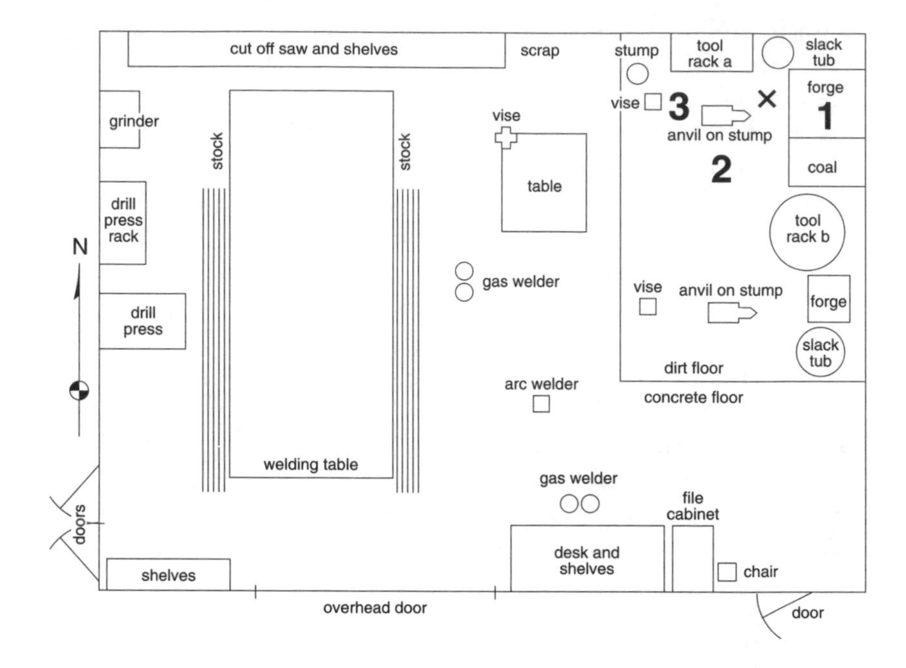

Figure 3.1. Diagram of Santa Fe shop in which CK worked as an apprentice in 1976 and 1977. Numbered features are the forge, anvil, and vise, which constitute the essential tools for forge work and define the primary work space. The X marks the center of this primary area. Not to scale.

trained. A diagram of the shop appears in Figure 3.1. The numbered features are the forge, anvil, and vise, which constitute the fixed elements of a tool inventory designed for the transformation of hot iron. The area bounded by these three elements is conceived of as the core location for the activity at hand; the remainder of the smith's tool inventory is located in varying degrees of proximity to it. Clusters of additional tools are located in the vicinity of the fixed features where they are commonly used. For illustrations of an anvil and a forge, see Figures 2.1 and 2.7. Figure 3.2 illustrates the standard post vise found in most blacksmith's shops.

The forge of the Santa Fe shop, labeled 1 in Figure 3.1, is the main shop forge used by the owner-smith. A slack tub and sprin-

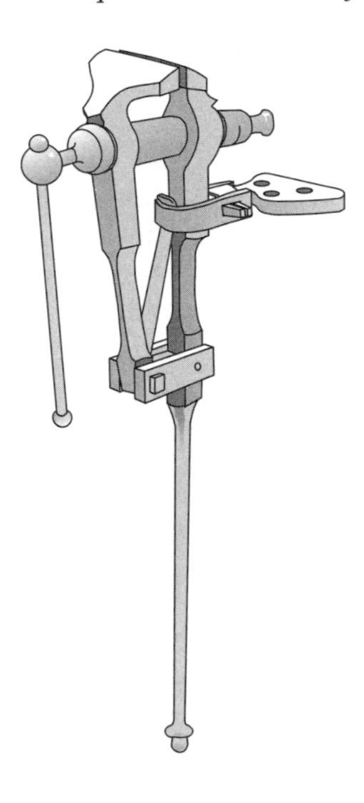

Figure 3.2. Standard post vise found in most blacksmiths' shops. Redrawn with the author's permission from Jack Andrews, *Edge of the Anvil* (1977), p. 25 (Emmaus, PA: Rodale Press). © 1977 Jack Andrews.

kling can for cooling iron and adjusting the fire stands just behind the forge. Fire tools including pokers, a spoon for removing clinker, and frequently used tongs are kept on the forge, where they are accessible for maintaining the fire or holding the iron while it is heated and carrying it to the anvil. Tool rack (b), located between the two forges, holds additional tongs and forging tools for use at the anvil, including punches, chisels, slitters, and bottom tools such as swages, bicks, and hardies, which fit in the hardie hole of the anvil. The tongs both at the forge and on the rack hang suspended from a bar, while punches, chisels, slitters, and bottom tools – all single-component tools – fit into pockets recessed into the table top of tool rack (b).

A variety of tools are kept at the anvil labeled 2 in Figure 3.1. In this shop frequently used hammers are placed on the floor around the anvil, standing on their heads with their handles upright and readily accessible. Hanging from the stump are a brush for removing scale from the hot metal, a hot rasp for filing off burrs, a bick for making bends smaller than that allowed by the horn of the anvil, a rivet header for upsetting rivets and backing their heads, and a hardie for cutting stock. Other small tools such as slitters, drifts, punches, and chisels might be kept on the stump for the duration of a job. "In effect the blacksmith's anvil is his workbench and as such has an endless variety of ancillary tools needed for shaping, cutting, and bending hot iron" (Bealer 1976:75; see also Vlach 1981). The most frequently used of these tools are regularly kept at the anvil.

Tool rack (a) holds hammers that hang from their heads on a table edge. These hammers are less frequently used than those kept at hand at the anvil. Handled hot tools with transverse heads that permit hanging are also located here, and sledge hammers are nearby.

On the vise posts are hung different-sized bending forks and jigs and twisting tools such as a pipe wrench, a crescent wrench, and pliers. A hacksaw is also kept here.

The drill press rack-table in Figure 3.1 holds bits, vise grips, clamps, squares, an electric drill, extension cords, and a face mask – all items typically used with the adjacent drill press or the grinder and cutoff saw that stand nearby. Cutting oil for drilling is typically kept on the drill press itself.

On a rack at the large acetylene gas welder, arc-welding helmets are hung. The welding cart, which holds tanks of acetylene and oxygen, is usually south of the small table, where strikers for lighting the torch are also usually kept. A graphite block, which provides a surface for welding, is kept near the torch. Typically a grinder is kept on the small table for use in cleaning up welds.

Stock for forging is kept on either side of the large welding table. Atop this table are spacers, some clamps, a wire brush, vise grips, a file, measuring tapes, and cleanup tools.

Near one door to the shop stand a desk, filing cabinet, and shelves where paperwork is kept. Chalk for drawing and a first-aid kit are stored here. Installation tools and painting tools are also stored here along with assorted miscellaneous items.

This locational arrangement of significant tools in one Santa Fe blacksmith's shop reflects the smith's knowledge and practice of his craft. The crucial spatial relationship among the three primary fixed features of the shop – anvil, forge, and vise – reflects the basic transformative principles discussed in Chapter 2 and the manifestation of those principles in practice. The triangle of anvil, forge, and vise must be placed so that the heated iron can be carried with a minimum of wasted time to the anvil or vise for working. Other tools are located around the major fixed features on the basis of their general functional associations, frequency of use, current use, and shape as it affords ways of storing.

The Illinois shop

A comparative look at CK's current shop, established in a central Illinois building originally constructed as a blacksmith shop in the 1870s, illustrates the same principles and associations in its basic organization. CK himself devised the present shop organization. But the principles involved appear to have been relevant for the original occupant, who established the location of the forge and trip-hammer, the planks for tool racks (b), (c), and (d), and the nails for hanging things on ceiling joists, all of which are utilized by CK in the contemporary shop.

The arrangement of the primary fixed features in this shop facilitates "working hot." As is the case in the Santa Fe shop, these tools include forge, anvil, and vise. The Illinois shop adds a trip-hammer, labeled 4 in Figure 3.3, to the basic triad of the Santa Fe shop, providing an alternative mechanism for increasing the efficiency with which some forging tasks can be accomplished. The trip-hammer, illustrated in Figure 3.4, is a mechanized hammer that uses a mechanical ram to forge iron. Trip-hammers were common historically in blacksmith shops in the Midwest, where they facilitated the sharpening of farm equipment such as plowshares. They have been preserved in this region in many artist-blacksmith shops.

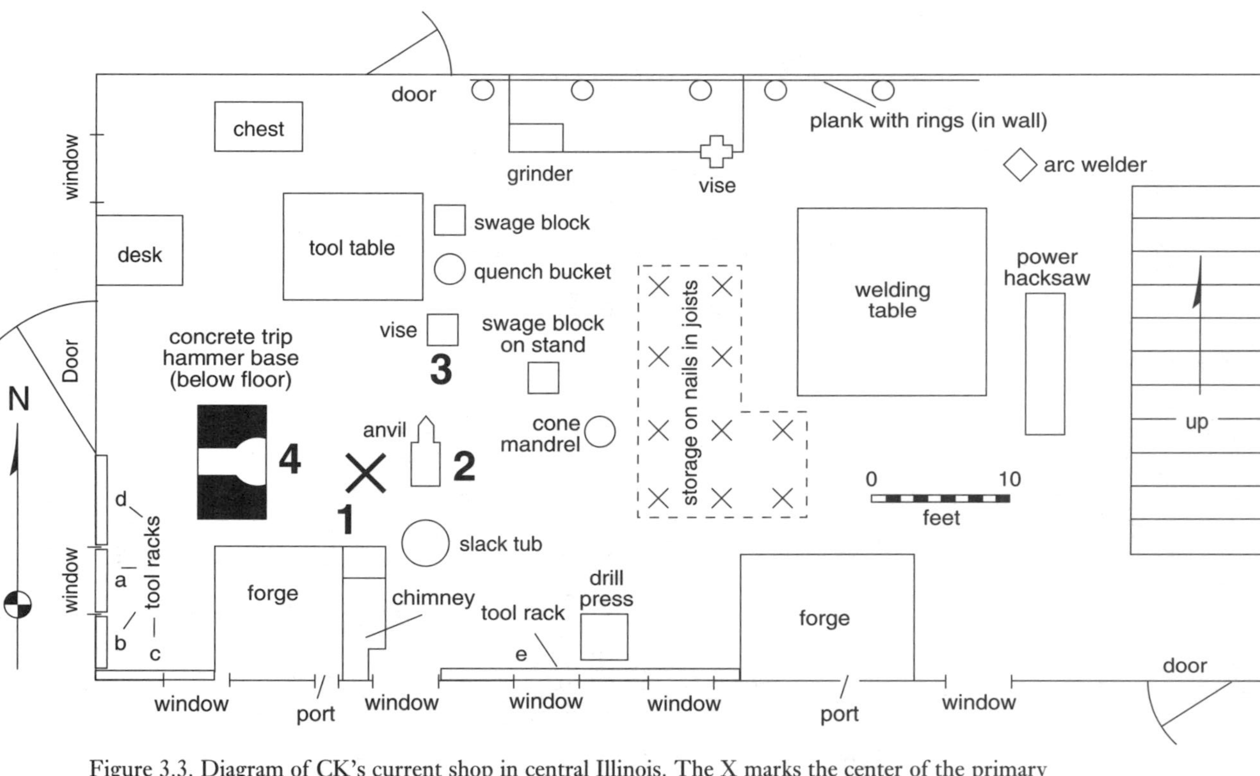

Figure 3.3. Diagram of CK's current shop in central Illinois. The X marks the center of the primary work space.

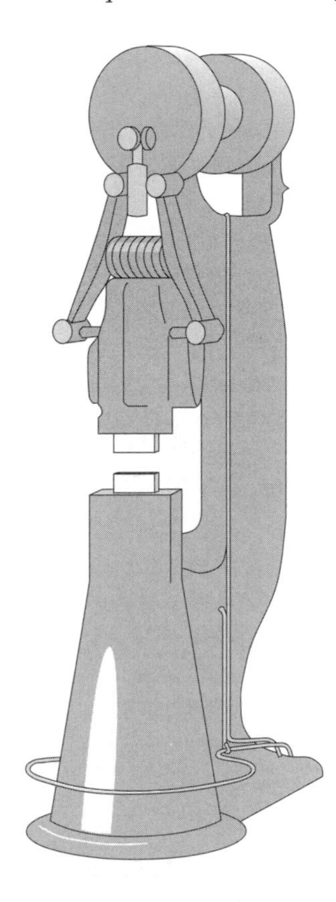

Figure 3.4. Fifty-pound trip-hammer. Height, approximately 6 feet.

The ram in this case weighs 50 pounds, although those found in artist-blacksmith shops may range from 25 to perhaps 125 pounds. The speed with which the trip-hammer operates can be controlled by the smith. It can be used with a diverse set of dies. The mechanical control is a circular treadle that allows the smith to move around the base of the trip-hammer while working. The vibrations accompanying the operation of this equipment and the blacksmith's need for freedom of movement preclude tool storage in the immediate vicinity.

Once again these immovable objects enclose a space that consti-

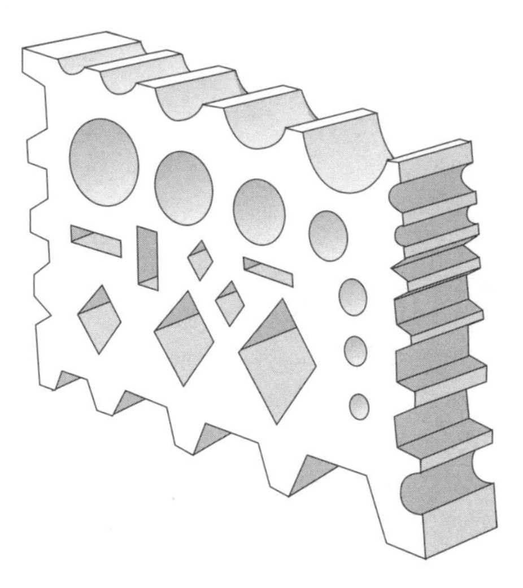

Figure 3.5. Common swage block. Approximate size, 15 inches by 13 inches by 6 inches. Redrawn with the author's permission from Jack Andrews, *Edge of the Anvil* (1977), p. 28 (Emmaus, PA: Rodale Press). © 1977 Jack Andrews.

tutes the hub of the shop and the tools used there are located nearby physically and visually. The tools kept on the forge are the pokers, a shovel, and tongs to fit the stock currently being used. The slack tub with its sprinkling can sits next to the forge. Hanging from nails driven into the brick chimney of the forge are a wire brush for removing scale and a spoon for applying flux to welds. The fluxes are kept in cans sitting on a projection of the forge below the spoon and brush.

At the location labeled 2 in Figure 3.3, the most often used hammers stand on the floor leaning against the right side of the stump under the anvil. Other tools that happen to be in use, such as a hot cut or a flatter, stand to the left of the stump. The placement of the tools on the floor is determined by the hand in which they will be gripped. A hot rasp is held by a loop attached to the stump behind the anvil. A hardie and a hold-down used to secure stock to the face of the anvil are kept at the stump along with any other anvil tools being used, such as a bick or a bolster (a plate pierced with

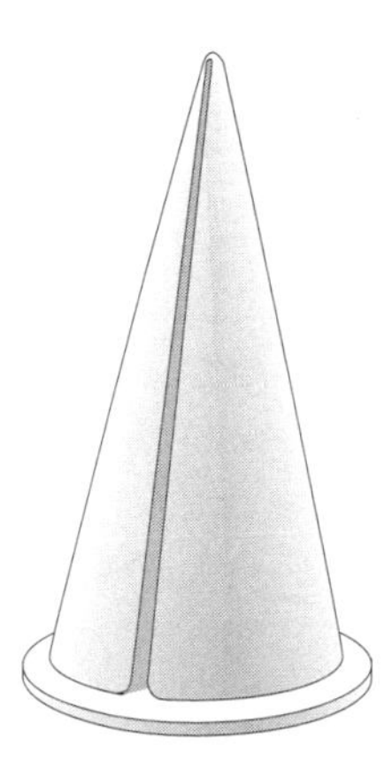

Figure 3.6. Cone mandrel. Approximate height, 4 feet. Redrawn with the author's permission from Jack Andrews, *Edge of the Anvil* (1977), p. 29 (Emmaus, PA: Rodale Press). © 1977 Jack Andrews.

holes of differing sizes that supports a work piece when it is being punched). Sitting on the stump next to the anvil is a center punch made from a jackhammer bit, used to mark locations on the iron so that the marks will be visible when the metal is hot. A measuring rule and welders' talc are located there as well as another wire brush, coarser than the one hanging on the chimney.

Within a step or two of the anvil are a swage block on a stand and a cone mandrel, both of which are used for shaping iron while it is hot. Both are fairly heavy but can be moved nearer to the forge if necessary. Figures 3.5 and 3.6 illustrate this equipment.

Racks (a), (b), and (c) near the forge hold ancillary tools. Rack (a) supports frequently used hammers; rack (b) holds tongs, examples of which are depicted in Figure 3.7; and rack (c) stores rarely used hammers and a pair of anvil shears. Tool rack (d) is attached to the wall at about waist height and holds screwdrivers, pliers, files and rasps, bicks, cold chisels, and other implements used on cold metal or wood. Tool rack (e), at some remove from the primary work area, consists of two rods with a narrow gap between them. This rack holds tools occasionally or selectively involved in work in the primary space. Various handled implements are stored here, including

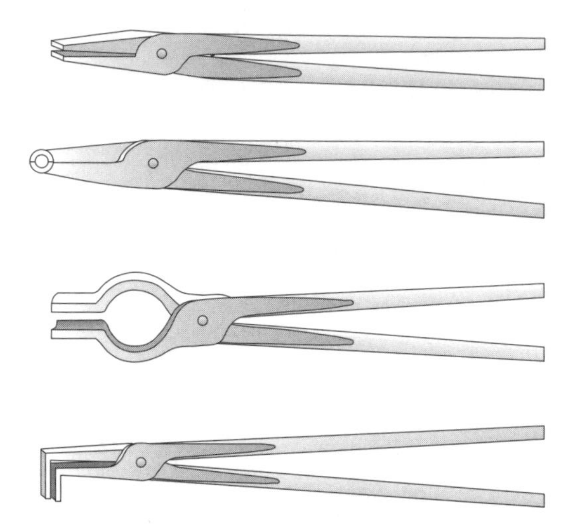

Figure 3.7. Variations in the grasping jaws of tongs are seemingly endless. Here four possibilities are illustrated.

hot and cold cuts, top swages, top fullers, punches, flatters, and set hammers, all of which can be suspended from their heads with their handles hanging down between the iron rods. Examples of the tools stored in rack (e) are depicted in Figure 3.8.

The vise (see Figure 3.2), which completes the primary grouping of four fixed features, is labeled 3 in Figure 3.3. Mounted on a post, the vise is located adjacent to a tool table that has a top surface, a shelf below, and rails around three sides. On top of the table are kept various hot tools like punches, slitters, and drifts, which stand on end in several large cans. Twisting wrenches and hinge-bending jigs used in the vise and an extra trip-hammer die lie on the top of the table as well. On the rails hang tongs that are never or rarely used but that came with the shop, as well as some rarely used spring tools. On the lower shelf is a wide variety of hot tools including bottom swages, bottom fullers, and others used at or on the anvil as well as those used under the trip-hammer.

In the nearby tool chest taps and dies used in the vise are kept, as are some infrequently employed items like buffing wheels and

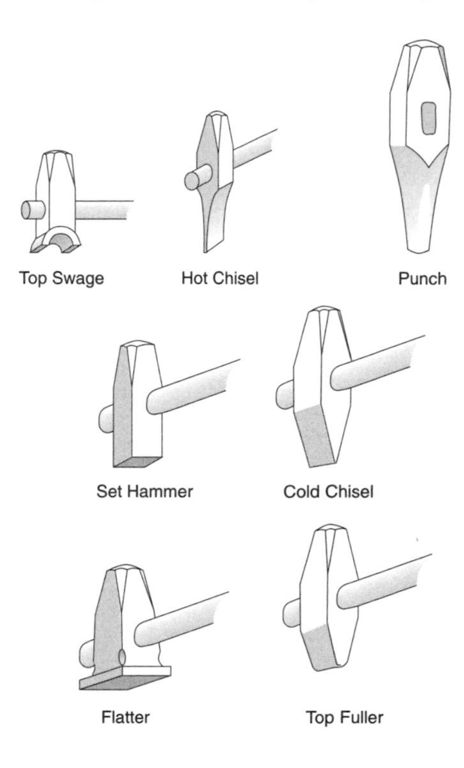

Figure 3.8. Selected handled implements used as needed for forging tasks.

compound. At the north door sits a stump on which a large and a small swage block are kept; another swage block sits close by on a stand. Near them is a bucket that contains oil used as a hardening quench for steel. These all can be moved to the forge when necessary.

Cutting oil and a wrench for adjusting the drill press table sit on an adjacent windowsill. Bits, a drill press vise, first-aid supplies, monkey tools, and some obsolete pipe-threading equipment are found on the second forge next to the heating stove.

A workbench with an additional vise fixed to the east end is located against the north wall of the shop. This vise is well lighted and is used for holding stock to be modified while cold, as in decorative filing. A bench grinder sits on the west end of the workbench and hearing protectors and a face shield hang on nails near it. Abrasive strips hang from nails in a joist above and to one side of the grinder. On a shelf under the bench the angle grinders are kept, along with some toolboxes containing miscellaneous small items and nuts and bolts. Compasses, scribes, squares, and a hand drill hang on the wall above the bench.

New stock and longer cutoffs are kept in a rack on the north wall while shorter cutoffs and scrap are stored in the corner.

A measuring tape, torch tips, and arc-welding electrodes are kept on the welding table, as are several heavy objects used to hold work pieces in place during welding or brazing operations. Also present are a wire brush, a chipping hammer, and a cold chisel, all of which are used for cleaning up welds after they are completed. A welding helmet, cap, and gloves hang on the wall near the arc welder.

Hanging from nails in ceiling joists about 10 feet above the center of the shop are clamps that can be used on the welding table, drill press, or workbench. Hanging nearby are various patterns and bending jigs. These tools can be retrieved from the ceiling by means of a pole with a hook on its end. A similar device was part of the original equipment of the shop.

As with the Santa Fe shop, the primary fixed features – in this case, forge, anvil, vise, and trip-hammer – are configured to enable the transformation of iron effectively while it is in the plastic state that results from being heated. This space is intensely personal, constituting the setting of creative production (Lave 1988). CK reports a sense of unease when visitors enter the area. It is important to note that this focal area of the work space could not be identified from the physical features alone without an understanding of the underlying principles of the activity. There are other locations surrounded by tools or equipment that do not have the same technological and symbolic significance as the area bounded by the forge, anvil, and vise. For example, were this a fabricator's shop, the primary work area would be located at and around the welding equipment. Recognition of the importance of the primary forging area is possible only if the significance of the pieces of equipment that bound the space is understood.

The location of other tools and equipment in the shop reflects the blacksmith's knowledge of their general functional relationships. All the tools used for hot work are found in the west third of the shop near the forge and are stored in racks or on surfaces as their particular shapes allow. Tools frequently used with the anvil, trip-hammer, or vise are located within reach of these features. Less frequently needed items are conveniently stored for easy retrieval when needed for ongoing work. Other functional associations are also in evidence. The items used with the grinders are located adjacent to, above, or below the grinders themselves. Things necessary for the operation of the drill press are located near it; the same is true of the items associated with the welders.

The basic approach of forging hot and the knowledge of general functional associations among tools are the primary factors in establishing standard locations for the implements of this shop. Opportunities for storage provided by knowledge of the forms of tools are secondarily taken advantage of as functional requirements permit. For example, given the shapes of many of the implements found in tool rack (d) in CK's shop, such as screwdrivers, cold chisels, and pin punches, these could equally well be stored on the tool table adjacent to the vise. But if they were stored there, along with similarly shaped hot tools, they might be seized by mistake in a moment of haste, and they would be damaged if applied to hot iron. Anticipation of the potential for such errors, which is based on a knowledge of the requirements of hot versus cold work, leads the smith to locate tools for cold work in a distinct rack that makes them easily accessible for activities such as assembly and disassembly but safe from an erroneous application.

Comparison of the two shops

The arrangements of the tools in the Santa Fe and Illinois shops are actively constructed by the blacksmith based on principles of transformation, working hot, and working freehand. Both these defining principles and the functional associations of tools based on these principles constitute aspects of a stock of knowledge organized for accomplishment in forging. The logic of the arrangement of tools is functional both in general and in more specific terms. On the basis of such knowledge, the blacksmith constructs expectations for future projects and these expectations become materialized in the arrangements of his implements.

The fundamental organization of the shop follows from the principles already mentioned, coupled with the requirements of accessibility and visibility discussed earlier. Morphological characteristics of particular groups of tools afford various means of storage, but these must be consistent with the orienting principles and requirements that allow an activity to proceed quickly and smoothly. Those tools with similar functional potential and similar shapes are grouped together so as to be easily seen and, for the most frequently used tools, also easily reached from the core area in which the smith operates. The organization thus derived is a general-purpose scheme for facilitating a range of activities likely to be required in the blacksmith's work (see Suchman 1987).

An inventory of tools: implications for procedural aspects of the stock of knowledge

These dimensions of a blacksmith's stock of knowledge, evidenced in the arrangement of a shop, set the stage for production. However, the organization of a shop in itself reveals little in specific terms of the procedures available to a smith. Such procedures are the frequent topic of conversation among smiths both in interactions and in their literature. Conversations regarding procedures often focus on the tools involved. Taking our cue here from these conversations, we use the tool inventories in the blacksmith shops depicted in Figures 3.1 and 3.3 as an entree into the details of the procedural component of the blacksmith's stock of knowledge. By focusing on particular tools we create an opportunity for CK to reflect on his own procedural knowledge and on that of the Santa Fe smiths from whom he learned his craft.

The following discussion of procedures can only be illustrative. As with other aspects of the stock of knowledge, what the blacksmith knows of procedures is extensive and potentially grows and transforms with every instance of production.

Further complicating the issue is the fact that the most frequently used tools, such as the anvil and hammers, are general-purpose tools utilized in diverse procedures rather than specialized for specific techniques. Almost every job undertaken in a blacksmith's shop employs these implements in one way or another. We will note some of the primary techniques made possible by these tools, but the technical variations are endless and the opportunity for creating novel possibilities is always present. Since new ideas for solutions to problems often result from the visual recognition of the potential of a feature of a tool commonly used in another way, visual contact between the smith and his suite of implements is as important as the ability to retrieve the tool physically. Blacksmiths tend to conceive of even the simplest of tools as multidimensional objects whose different parts can be used in different ways. Thus the tip of a punch makes a hole of a certain diameter, but the tapered shank of the punch can also expand that or another hole if it is driven into stock. Or, as one of CK's mentors instructed him, different parts of the face of a hammer can be used to produce different results. "Look at [the] hammer face to see what part you want to hit with. [The] edges and full face can all do different things" (C. Keller 1976: Feb. 25). In addition, while a given tool may be used for a given technique – a punch for punching, a hot cut for cutting – each procedure can be done incompletely to produce a different end result. An incomplete cut may become a decorative groove, or an incompletely punched hole may form an ornamental depression.

Procedural knowledge is complicated as well by the distinction between techniques and recipes. Any given tool may be involved in a *technique* for performing a particular operation. Accomplishing that technique will also usually be only one step in the production of a particular end product. Punches, slitters, and drifts make holes. But what the hole is in, where it is located, and the way the hole is related to other elements often involve a larger *recipe* for the production of a given product. A given tool inventory, therefore, when conceptualized by a working blacksmith as an inventory, is likely to index both the techniques and the recipes of past and anticipated projects.

Frequently used hammers, the forge, anvil, and vise constitute the required equipment for primary transformative forging processes. Specific procedural knowledge distinguishes among the many transformations enabled by this basic equipment. For example, changes of shape that can be accomplished include *drawing out* – increasing the length and/or width of a section of iron by decreasing thickness; *upsetting* – increasing the width and thickness of a section of iron by decreasing its length; and *bending* – changing the line, angle, or curvature of a section of iron. These transformations of a length of iron are accomplished using general-purpose tools and are referred to by the general term *forging*.

Techniques for accomplishing these transformations require distinctive applications of the same implements. The stock of knowledge for a blacksmith will include a variety of forging techniques to draw out, upset, and bend iron. We will use a set of possibilities for drawing out a length of iron to illustrate this characteristic diversity.

One procedure for drawing out is utilized by CK for creating a tapered segment of iron rod by using the forge, anvil, and handheld hammers as follows. A round rod is squared up to create flat surfaces for forging and then hammered to a point by laying the end of the squared rod just at the far edge of the anvil with the length of the rod running perpendicular to the anvil's axis. By raising the end of the rod toward the smith and hammering at the far end where it is placed at the edge of the anvil, the rod can be pointed. Then, for further pointing and tapering, the tip of the rod is extended over the edge and the hammer face is worked parallel to the axis of the rod. When the desired taper is achieved, the corners of the squared rod should be cut off, that is, hammered out to re-create a rounded tapered section. When cutting off the corners the rod must be rotated back and forth, reversing direction between consecutive hammer blows, to avoid giving a spiral twist to the rod (C. Keller 1976: Feb. 27, March 16–29).

This procedure produces a length of tapered rod ending in a point. A similar but distinctive technique for drawing out is used by CK in working with flat rather than square stock. In this case, the stock is held flat on the anvil and hammer blows are concentrated at the end of the piece. This thins, stretches, and widens the stock; the

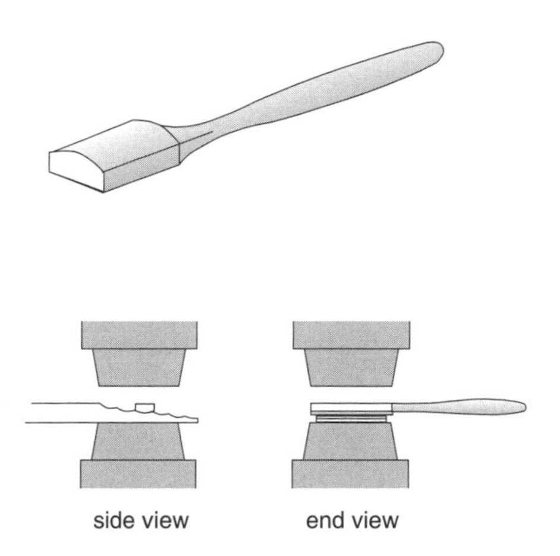

Figure 3.9. Drawing out iron by interposing a handled tool between the dies of the trip-hammer.

taper is produced by using fewer and lighter blows as the work moves away from the tip.

These procedures are standard techniques for drawing out utilizing only general-purpose equipment or tools of risk typical of blacksmithing: forge, anvil, and handheld hammers. The techniques are common to the smiths working in both the Illinois and Santa Fe shops and, in fact, are probably shared by most contemporary artist-blacksmiths. If we look beyond the basic triad to the fourth primary tool in the Illinois shop, we find techniques for drawing out with the trip-hammer. For example, CK commonly uses two techniques for drawing out stock with his trip-hammer. The first involves using a top die, called a drawing die, which has a markedly curved face. This reduces the thickness of the metal very quickly but leaves a slightly irregular surface, which may have to be removed with a hand hammer. This technique also produces a basically flattened section of iron of roughly uniform thickness. If a sharp taper or point on the work is desired, an alternative technique must be employed. Drawing dies cannot be used to achieve these ends. The alternative, illustrated in Figure 3.9, is to use a flat-faced top die and a handled tool with one curved face and one flat face that is interposed between the moving top die and the work. If the flat face of the tool is placed uppermost, the curved face of the tool works much like a drawing die. However, if the curved face is used uppermost it is possible to shift the angle of the flat lower face of the tool on the work and to produce sharp tapers or points.

These techniques for drawing out with a trip-hammer take advantage of the force available from the weighted ram to change the dimensions of raw material and shape the work by using variously shaped dies. Because of the mechanical design of the trip-hammer it is not possible to vary the angle of the forging blow. The use of dies therefore enables the accomplishment of desired ends that are alternatively achieved through procedures of drawing out by hand by relying on the precise angle of the work relative to the anvil and the angle at which the blacksmith strikes the iron in the process of forging.

Turning to more specialized tools, CK's reflections include additional techniques for drawing out iron stock. For example, tools referred to as fullers appear prominently in the Illinois shop depicted in Figure 3.3. These can be distinguished as top and bottom fullers. A bottom fuller is a loaf-shaped tool with a shank that is placed in the hardie hole of the anvil. Fullers of different sizes have heads the radii of which are different. Using a fuller to draw out a section of iron tends to allow the blacksmith to move more iron in fewer heats than would be possible with handheld hammers. This procedure involves inserting a bottom fuller into the hardie hole of the anvil and placing a hot iron rod or bar perpendicular to the axis of the fuller. Figure 3.10 illustrates the process. The work is hammered with the top fuller or handheld hammer to create a pair of depressions from the top and bottom tools. The work may be repeatedly lifted and placed on the fuller again at points successively removed from the original placement and hammered to create a series of regular depressions in the metal. After one or more pairs of depressions have been forged into the work, the iron is placed flat on the anvil and the intermediate ridges hammered down to produce a narrower rod of greater length than the original.

The blacksmith's tools, both general-purpose and more restric-

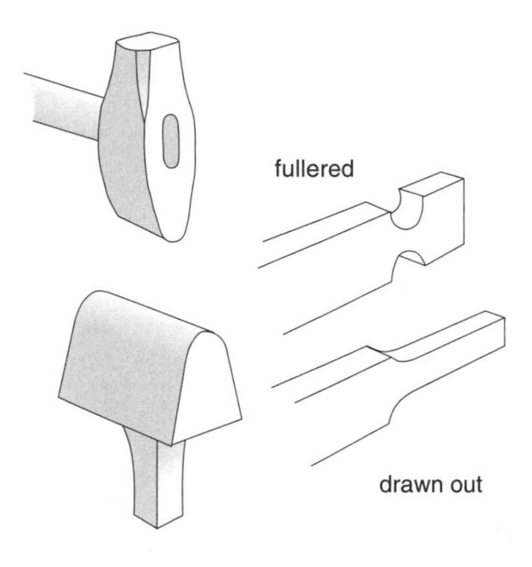

Figure 3.10. Fullering as a technique for drawing out. Redrawn with the author's permission from Jack Andrews, *Edge of the Anvil* (1977), p. 27 (Emmaus, PA: Rodale Press). © 1977 Jack Andrews.

tive, index his knowledge of working procedures. Choices among different procedures for accomplishing the same end are made on the basis of a variety of factors. These may include such things as the number of pieces to be drawn out, the size of the stock, the length of the taper, and time constraints on a job.

Fundamental forging techniques such as those just discussed may be combined to solve a particular problem or to produce a particular form. Such standard combinations of techniques become recipes in the smith's stock of knowledge and can be put to use when the need arises.

For example, in the Santa Fe shop a fishtail scroll (see Figure 4.6, p. 99) was often used as a decorative element. The smiths working in the shop produced this form using a recipe that combined techniques for drawing out and bending. The recipe is roughly as follows. The end of a bar or rod of sufficient length for the project at hand is heated and placed on the anvil and drawn out (by the second technique mentioned above) so that the stock tapers gently from its original thickness to the desired thickness at its end. Drawing out thins the stock and lengthens it, but it also increases its width slightly. In the case of the fishtail scroll the increased width may be allowed to remain and the edges forged lightly to create a flared line. Next the stock is heated, placed flat on the anvil, and forged to remove irregularities on the upper and lower surfaces. The final operation before returning the work to the fire for the next step is to check for straightness and correct if necessary. For the scroll to form smoothly it is crucial that the tapers of both the plan form and the cross-section be uniform.

When the work has been reheated sufficiently, it is taken from the fire and any scale removed by wire brushing. The scale must be removed at this point because after the scroll is formed its curvature will make brushing difficult or impossible. After brushing, the work is placed on the anvil with the tip extending slightly beyond the edge. The tip is tapped lightly; then the piece is advanced a fraction of an inch and tapped again. This continues until the desired curve is produced, or until the work is cold, or until the downward curving tip contacts the side of the anvil. When the bending is halted, the curve is inspected from the side and the shape corrected if necessary. The maximum curve achievable at this stage is a semicircle. To increase the size of the scroll the portion of the bar immediately behind the curve is heated and the outside of the curve already formed is struck lightly with the hammer to continue rolling up the scroll. Alternatively, if the size of the stock requires it, the horn of the anvil can be used as a fulcrum and the work bent over it. Whatever the technique chosen at this point, the bending is continued with constant checking of the piece from the side and correcting as necessary until the desired form of the scroll is achieved.

Recipes for a large variety of basic products such as the fishtail scroll and more complicated projects, such as a gate incorporating scrolls as design elements, are part of a smith's stock of knowledge and can be elicited ad infinitum when one asks about the potentials of various tools. Like techniques, recipes are knowledge organized for doing. It is this procedural knowledge constructed in production that the smith then draws upon to continue to make the products of his craft.

The importance of particular recipes and techniques for a given blacksmith is integrally tied to the smith's talents and abilities and to the products he makes as well as to the tools of his trade. If we return momentarily to the procedure for drawing out by fullering, the significance of a smith's personal characteristics for the salience of a particular technique can be clarified. Drawing out by fullering is a technique CK learned as an apprentice in Santa Fe, not from the smith with whom he worked regularly but from a smith in a neighboring shop, who suggested the technique should reduce the time required to draw out a section of iron by two-thirds (C. Keller 1976: April 6). The primary smith in this case said that he never bothered with fullering; commenting further, he suggested that "maybe" he was stronger than either CK or the neighboring smith (C. Keller 1976: April 6), an accurate observation. The continued significance to CK of the technique of drawing out by fullering is partially a factor of his personal strength.

The nature of the artifacts of a blacksmith's production also influences the selection of tools for his work and thereby affects the relative importance of particular techniques and recipes for him. Recall that the Santa Fe shop is organized around the triad of forge, anvil, and vise, while the Illinois shop incorporates a trip-hammer into the basic equipment. In both cases the shops are set up to facilitate the transformation of iron while it is hot. However, the work done in the Santa Fe shop is primarily ornamental and architectural. The products at issue include railings, gates, and window grills, as well as household accessories such as fireplace screens and tools, kitchen utensil racks, and other hardware. Production of such artifacts typically takes advantage of raw material available in dimensions approximating those needed for the end product desired. Changes in shape required by the ornamental designs typically involve drawing out and tapering, twisting, bending, and scrollwork. Because the dimensions of the raw material approximate the dimensions of the completed work, the transformations required in production do not usually entail significant movement of a considerable mass of iron. The smith operating the Santa Fe shop found these transformations could be produced effectively with handheld hammers.

The work done in CK's shop, on the other hand, is primarily replication of colonial-period tools. Production of early American tools does require significant changes in the shape of raw material of considerable mass. For example, the stock used in producing a full-sized Kentucky axe consists of three pieces of ½-inch by 4-inch stock and one piece of ¼-inch by 1-inch stock. These are laminated by forge welding and the thickness of the resulting 1½-inch-thick billet is reduced by half. Typical dimensions of the raw material used for ornamental work in the Santa Fe shop is by contrast ½ inch square and its cross-section was usually changed only by tapering. As a result of CK's need to transform material of larger dimensions, a trip-hammer became a useful addition to the basic forging triad in the Illinois shop and drawing out by means of the trip-hammer became an important technique for him.

We can extend the discussion of procedural knowledge beyond the primary changes of shape created in forging by focusing on other tools. For example, in the Illinois shop items including a wire brush and fluxes are part of the inventory of implements kept near the forge. These are involved in forge welding, a technique involving heating pieces to be welded to a very high temperature and then joining them with light hammer blows.

By all accounts forge welding is a high-risk procedure with much lore surrounding it. While there is some physical danger to artist-blacksmiths associated with working with open fires and hot metal, the risk referred to here is the risk of failure. For example, blacksmiths report that "welding is magic and can't be done" or "you can't get a good weld with your teeth in" or "the older the smith the darker the welding heat." Forge welding is one of those procedures involving tools of risk, as discussed in Chapter 2. The work done in CK's shop, primarily the replication of colonial-period tools, requires forge welding, the only traditionally available technique for merging two pieces of stock. The blacksmithing technology of 18th- and 19th-century America relied on this technique, particularly for welding together iron and steel to form cutting edges. On the other hand, forge welding is not practicable for joining elements of the large architectural products produced in the Santa Fe shop. Both the size and the design complexity of these products typically preclude this process as an effective construction technique. Thus procedural knowledge of forge welding is an important component of CK's technical repertoire but insignificant as a practice for his Santa Fe mentor, who works exclusively with modern techniques of gas and arc welding.

Hardening and tempering are also processes required for the production of tools, but not for contemporary architectural work. If you ask about the quench bucket that can be found in CK's shop but not in the Santa Fe shop, the answer will be about procedures for hardening and tempering. Heat treating, or hardening and tempering high-carbon steel, requires two stages. First, steel is heated to about 1,500° F, which is its critical temperature – the temperature at which its crystalline structure changes. The steel is then quenched in oil or water. The blacksmith typically determines when the steel has reached this temperature by applying a magnet to the work. When the work has lost the property of magnetism, it has reached the critical temperature and is immediately quenched. This process hardens the steel, which leaves it in an extremely brittle state. To make it useful it must be subsequently tempered, which requires polishing the surface of the work and reheating it this time to a temperature of between 420° and 630° F (Andrews 1977). The temperature of heated steel is apparent in the color of its polished surface. The tempering colors range from light yellow through brown to blue, each color reflecting a particular, relatively narrow, temperature range (see Figure 2.6). Temperatures in the range between 420° and 630° impart characteristics of relative resilience and sharpenability to the steel. The blacksmith heats the iron to a color chosen for the qualities desired and allows it to cool.

An inventory of tools in a shop now can be seen to reflect the procedures of past and anticipated projects in which that inventory has been or might be instrumental. A blacksmith's tools are typically accumulated with specific productive tasks in mind; in utilizing his tools the smith relies upon his underlying conception of previous or projected accomplishment. (See Hunn 1985 for discussion of the related notion of activity signature.)

This knowledge constitutes an encyclopedic foundation both for reproducing techniques and for innovating in practice. This rich foundation of schematic knowledge is governed by the principles of artist-blacksmithing. Potentials for transformation, distinctions between hot and cold procedures and equipment, and the ability to work freehand are central to the sedimentation of schemata from practice. Yet the information represented by procedural knowledge is only guided, not determined, by the common principles. Procedural schemata are constructed and reproduced as the smith balances principles with commercial considerations in creating technical compromises that take advantage of the capacities of his tools and the arrangement of the work space (Gibson 1977; Norman 1988).

Discussion

The argument we wish to derive from this discussion applies to blacksmiths in general as well as practitioners of other activities. We have demonstrated that the physical arrangement of the work space reflects basic principles of a worker's stock of knowledge and indexes specific techniques and recipes for production.

The selection and placement of basic equipment and tools in a shop clearly manifests the actor's sense of what will be needed and where for carrying out a certain range of activities. The inventory and arrangement of tools reflect the anticipation of satisfying requirements that an actor, through the sedimentation of past experience, can expect to encounter in the process of his future activities.

Note that while the blacksmith's shop provides a structured arena for work, it is somewhat different from the arena recognized by Lave (1988) in her studies of grocery shopping. In the case of American shoppers, the context for activity is socially structured independently of the specific goals of any particular shopper. In the case of the blacksmith shop, however, it is typically a smith who organizes his own work space. It has been essential, as a result, to work with individual blacksmiths and their shops in establishing the features of significance and their relationship to the knowledge of smithing.

While the blacksmith has more responsibility than the shopper for constructing the arena of his activities, the mutually constitutive relationship between knowledge and practice is as evident in this work space as it is in the grocery store. A tension between individually variable dimensions of smithing and common premises accounts for shared elements of shop design and variations in inventories of tools and their locations within a shop. For example, functional and formal features of tool design leave significant degrees of freedom for individualized development of tool use that takes advantage of personal talents and characteristics and tool availability. In addition, given tools may be used similarly or in different ways by different smiths and different tools may be used by different smiths or by one smith at separate moments to achieve common end results or for distinctive purposes. Consequently, reconstructing specific procedural knowledge in the case of blacksmithing again requires intensive work with individual practitioners. Yet beyond the variations correlated with personal qualities and selected goals of production, the principles and activities of artist-blacksmithing provide for an integrity recognizable from one shop to another.

Independent evidence for the components of the stock of knowledge discussed above and their practical organization comes from manuals (Bealer 1976; Friese 1921; Drew 1915; Andrews 1977; Holmstrom 1900) and historical works (Bealer 1976; Nolan n.d.; Wylie 1990; Wigginton 1979). Typically included in these is a description of shop layout with some discussion of the plasticity of iron and steel when heated and the episodic character of blacksmithing work. Typical also is a discussion of the centrality of the forge and anvil to blacksmithing, with special emphasis often given to the vise as "the third hand of the smith" (Andrews 1977:25). Manuals also often discuss useful implements in terms of their functional utility and shapes, and present basic techniques of forging. Contemporary publications of artist-blacksmithing organizations *(The Anvil's Ring, The Blacksmith's Journal)* take these general principles for granted, building upon them in discussions of the production of new tools and products, innovative strategies, and difficult-to-acquire information.

Our analysis of the underlying conceptualization of forging activities as reflected in a shop is a reconstruction of significant components of a blacksmith's stock of knowledge, both principled and schematic. As previously sedimented from instances of activity and manifested in anticipations for future activity, these components of the stock of knowledge are organized for the efficacious accomplishment of diverse ends. This knowledge is general-purpose in the sense that it governs the range of activities of an artist-blacksmith, including the organization of his shop. We turn now to a preliminary examination of the situated use of this knowledge in actual contexts of production.

4 Constellations for action

An inventory of tools and their locations relative to one another yields a large set of open possibilities for production. The smith chooses and utilizes elements of this inventory and associated experiences and anticipations as he carries out projects and solves problems. The question then arises as to how these open possibilities are selectively accessed and employed for achieving a specific end. This returns us directly to the question of how someone makes something. Schutz (1971) has argued that a course of action is selected from among possible alternatives on the basis of problematic aspects of a project of the moment. Insights from phenomenology combined with the emphasis of activity theory on goal-oriented actions have led to our second hypothesis: that features of a task orient a blacksmith to selectively access both the mental reservoir provided by his stock of knowledge and the pool of material resources provided by his tool inventory and store of metal. This review of mental and material resources is undertaken in an effort to construct an image of a desired goal and procedures for its attainment.

We use *task* loosely here to refer to the features orienting production of an artifact or an intermediate stage in its manufacture. Within the framework given by the principles of transformation, thinking hot, and working freehand which characterize artist-blacksmithing, a task orientation guides the smith's review of possible outcomes, procedures, and plans. Anticipated problems that arise in consideration of what needs to be done to achieve a desired end may lead to the rejection of one or more alternatives and selection of others.

Ultimately a plan is developed in which the final goal is mentally outlined and a sequence of means to this desired end is more or less roughly conceptualized. We refer to this overriding notion of a goal for production as an *umbrella plan.* This plan is open to revision and reflection in practice, providing a conceptual framework for correction and accommodation to unanticipated problems. The umbrella plan, which will be discussed in detail in later chapters, is the smith's mental representation of an ultimate goal for production and his associated general overview of the step-by-step procedures required to attain that end.

The *step* referred to here is the smallest unit of activity typically attended to by a blacksmith. What is unified in the mind of the smith as a step has an immediate goal for transforming the metal. It requires that the smith be in the primary working area with materials in hand and tools in reach and entails a particular physical action or actions on his part. A step is developed as the means or a segment of the means for attaining an ultimately desired outcome. Where the activity planned for achieving a task at hand requires only a single step, the final outcome of the task and the immediate goal of a step in production will be the same. Where multiple episodic steps are involved, intermediate goals can be distinguished from the final outcome. Completion of a given step in production may require one or more heats. However, moving to a subsequent step in production is almost always marked by reheating the material in anticipation of a new procedure.

The notions of specific means selected for the accomplishment of each step in the production of an artifact guide the blacksmith in the choice of raw material and implements. The resultant configuration of ideas, implements, and materials we refer to as a *constellation.* These elements, mental and material, are equally critical. They are mutually constitutive of the constellation as a whole and are held together by a logic constructed from the goal orientation and principles of the smith. Production of a particular item entails the selection of tools and materials on the basis of this plan in anticipation of enacting envisioned procedural steps. We hypothesize that constellations constitute the enabling unit of productive activity (Dougherty and Keller 1982).

We find evidence to support this hypothesis in the rationale for continual selection, use, and relocation within the shop of tools and materials employed in productive activity. We will argue that selections of materials and implements are governed by the blacksmith's constructed notion of appropriate means derived from his stock of knowledge in light of his understanding of the problems to be solved in accomplishing a task at hand.

Whether an envisioned production is novel or routine, it is characterized by a sequence of steps enacted through the application of constellations. Each constellation constitutes a supposition specifying means – material and governing conceptual notions – for achieving a particular transformative step in production. Enacting the step then verifies the constellation through the achievement of an expected outcome, or fails, resulting in the need to rethink the procedures and entailed materials and implements. Routinely applied constellations may become steps in recipes for production of particular artifacts. In these cases a constellation has proved successful in repeated applications. Following proven procedures reduces the likelihood of failure for the smith. Nonetheless, revision of routinized constellations may occur as novel circumstances are encountered.

Our notion of constellation is perhaps unusual in encompassing mental and material components as mutually constitutive elements of the whole. Yet there are useful precedents in the literature.

The term *constellation* and some of the ideas associated with our use of it have been applied by others. Link (1975) utilized it to refer to the associated set of criteria employed by a Japanese cabinetmaker to choose among bundles of stock for a specific project. The criteria she identified include notions of the physical attributes of the wood as well as socioeconomic factors as understood by the cabinetmaker. Our notion broadens her conception to include actual material items enabling production. The term was also used by Barker (1968) to characterize the material and physical context that enables an activity to proceed. Our notion of constellation, at once narrower yet richer than that of Barker, refers to a conjunction of enabling ideas and physical components for accomplishment of a step in productive activity. We insist on the essential dialectic of reason and physical elements in constituting a constellation

These senses of the term *constellation* share a common feature in the focus on crucial arrangements of physical components enabling activity. However, the theoretical frameworks in which the notion is respectively developed and embedded are sufficiently distinct that a comparison based on the term alone is misleading. Our usage of *constellation* combines the mental components enabling activity as referred to by Link and the material entailments referred to by Barker.

Other studies of activity often construct a concept similar to our constellation. As discussed in Chapter 1, Lave (1988) has developed a framework for studying knowledge and practice that highlights the *arena* in which an activity takes place and the *setting* for specific actions. The setting shares with the constellation an essential dialectic quality. The actor comes to the task of choosing among physically present options (items on a supermarket shelf) prepared to take action in the appropriate contexts, and with an already formulated expectation of the outcome. This combination of physical setting and mental preconditions in her example of grocery shopping shares elements with the constellation that enables the blacksmith's production.

Another example can be drawn from the work of Hutchins (1993a) on navigation aboard U.S. Naval vessels. Here navigational team members use visual bearings to landmarks on shore to note the position of the ship on a chart. In contrast to the case we have developed for blacksmithing, both the mental and material elements of this navigational constellation are often distributed among actors rather than concentrated within and about an individual. The constellation here is analogous to the unit we find basic to blacksmithing in the mutually constitutive relations of tools and a conceptualization of a task, yet the issue of knowledge representation has the added dimension of distribution – often of overlapping components – among team members. The material elements likewise are distributed in line with subtasks for which different persons may be responsible.

The instances described by Lave and Hutchins share a common feature that contrasts with our delineation of a constellation. Lave's shopper must enact her anticipated purchase in an arena and a setting established by others, and Hutchins's navigational team members must accomplish their subtasks in the context of a ship that they did not design or construct. By contrast, in forging the artist-blacksmith is ultimately responsible for the design and procedural steps of a project, for identifying or producing the enabling tools and materials, and, as discussed in Chapter 3, for determining their general arrangement in the shop.

Let us return to the example at hand to explore more fully the constellation in blacksmithing. At the completion of a task the physical component of a constellation assembled for accomplishing an end is disassembled and the locational integrity of the tool inventory is reconstituted as tools and equipment are put away and the shop is cleaned up. Consequently, the constellations we identify are only accessible to study through their material elements during actual production sequences.

We are directed in our investigation of constellations by our informants themselves. During his apprenticeship in Santa Fe, CK was reminded to think about what he was doing by focusing on the means to attain a desired goal and the requisite tools and materials. Within the first 10 days of his apprenticeship, a version of the following comment appears four times in CK's daily notes: "As you are standing taking a nice, good heat . . . plan the next step. What tools do you need? Are they in reach?" (C. Keller 1976: March 8).

The importance of dividing a task at hand into goal-oriented segments is widely recognized in diverse approaches to understanding activity (D'Andrade 1992; Vlach 1981:13; Gatewood 1985; Lave 1988:106, 164; Lave and Wenger 1991; Lemonnier 1992; Leont'ev 1981; Schiffer 1992) and by those engaged in productive activity as well. Willie, subject of Douglas Harper's book *Working Knowledge*, emphasizes this when he says of his project to maintain a 1929 Model A Ford, "You start, you do one part, and when you get that done you do another. So it's done in sections. You don't try to do an overall job at one time. One thing at a time." In fact, a hierarchy of segmentations is typical of productive activity that is governed by an overarching project goal or task at hand and achieved by enacting one or more steps (see also Gatewood 1985).

As we have pointed out in Chapter 2, time is critical for the smith during a forging sequence and having the necessary tools at hand is

Figure 4.1. View to the south as seen from the primary work space in CK's Illinois shop. Forge is centrally located. Tool rack (e) is visible on the far left. Flux spoon, wire brush, and flux containers are on the chimney. Fire tools and tongs currently in use are to the right of the fire.

imperative if work is to be done effectively. As the passage quoted above from CK's notes indicates, as an activity is planned, the location of the required equipment must be checked visually. If the requisite tools are not within reach, the monitoring phase of heating the iron must be suspended and the tools made accessible. A glance at a tool may trigger an idea for production. This is part of the rationale for fully visible storage of tools and material discussed in Chapter 3. Figures 4.1–4.4 illustrate four views CK has of his shop and its inventory as he stands within the primary work area. The smith must draw upon the stock of knowledge and inventories of materials and tools as depicted in these figures, extracting and potentially rearranging information relevant to the anticipated actions.

Figure 4.2. View to the west as seen from the primary work space in CK's Illinois shop. Tool racks (a), (b), and (c) are partially visible, with the trip-hammer apparent in the foreground.

Constellations illustrated

Take, for example, the description of a job from CK's Santa Fe apprentice notes. These notes refer to the task of twisting a pair of brackets to hold a handrail, a task that was novel at the time the description was written but that became more routine.

> I used half-inch rod and heated and hammered the center so it was semi-square. I heated it again and quenched the end and clamped it in the vise. I grabbed the opposite end with a small pipe wrench and walked around to give the twist. It was a nice soft twist because of rounded shoulders from semi-squaring the rod. I reheated the rod and straightened it with a big wooden club on a stump. That way by using soft stuff the twist wasn't distorted. Quenching was to keep the vise from mashing the round part when it was clamped. Then I heated it again and bent it with a fork to a 90° angle [to finish the bracket]. (C. Keller 1976: March 8)

This description, typical of many in his notes, is organized in terms of a particular task at hand, the steps for production, and the

Figure 4.3. View to the north as seen from the primary work space in CK's Illinois shop. The vise, the tool table, a swage block, and a tool chest are visible.

associated raw materials and implements. The goal of the task at hand is depicted in Figure 4.5. To accomplish this goal one twists a length of iron and then bends the twisted segment to the desired final shape. It is the problematic aspects entailed in creating the twist and bending the rod that orient the blacksmith. *Semi-squaring, quenching, twisting, straightening,* and *bending* are the transformations to be achieved in this case. Each of these transformations in the production of the bracket involves a notion of specific means and ends and each is associated with a particular set of enabling implements. From the possibilities offered by the shop and the blacksmith's stock of knowledge, particular elements are selected in what becomes a series of constellations for production of the brackets.

The twisted rod is straightened, for example, with a wooden club

Figure 4.4. View to the east as seen from the primary work space in CK's Illinois shop. The anvil is directly ahead, with measuring and marking tools and a hardie resting on the stump. Two hammers are positioned on the ground leaning against the stump. A second swage block is visible on its stand. Welding table and cone mandrel are also apparent. Although not evident in this photograph, the stock is also visible from this perspective in storage racks just to the left and above the welding table.

and a stump. The governing procedure for the selection of these implements is the light tapping of the twisted iron to bring the full length into line without distorting the twisted segment. The wooden club and stump are selected as the facilitating tools in order to maintain the well-defined twist in the center of the bracket and insure that the iron is in line throughout its length. Relative softness is a critical factor in choosing the wooden club and the stump for straightening; by contrast, the hard-surfaced hammer and anvil would mash the fine edges of the already twisted section.

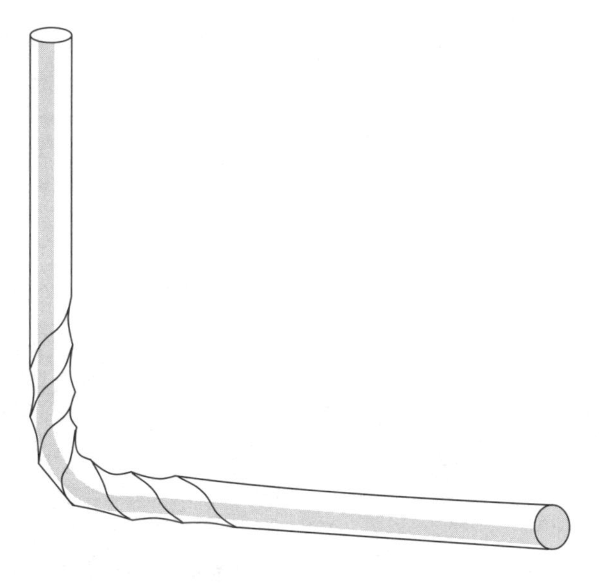

Figure 4.5. Twisted bracket for handrail.

The configuration of procedures, implements, and materials used for straightening is one example of a constellation. Another example comes from the technique of drawing out by fullering as one step in the production of a fishtail scroll. As discussed in Chapter 3, to make a fishtail scroll (see Figure 4.6) one lengthens and thins a bar of iron and flares the section at the end. With this as an orienting goal, fullering is one possible means for accomplishment. Anticipating this procedure, the blacksmith adjusts his fire to bring the raw material to a working heat and selects his tools. Implements included in the constellation are a fuller or the anvil horn to serve as the bottom tool for creating depressions, and a top fuller or hammer selected for weight and curvature of the face on the basis of the dimension of the raw material, the shape and size of the bottom fuller or horn, and the strength and working habits of the smith. A pair of tongs might be selected to match the dimensions of the iron rod and a wire brush would be placed near the anvil stump to be in reach for removing scale from the metal.

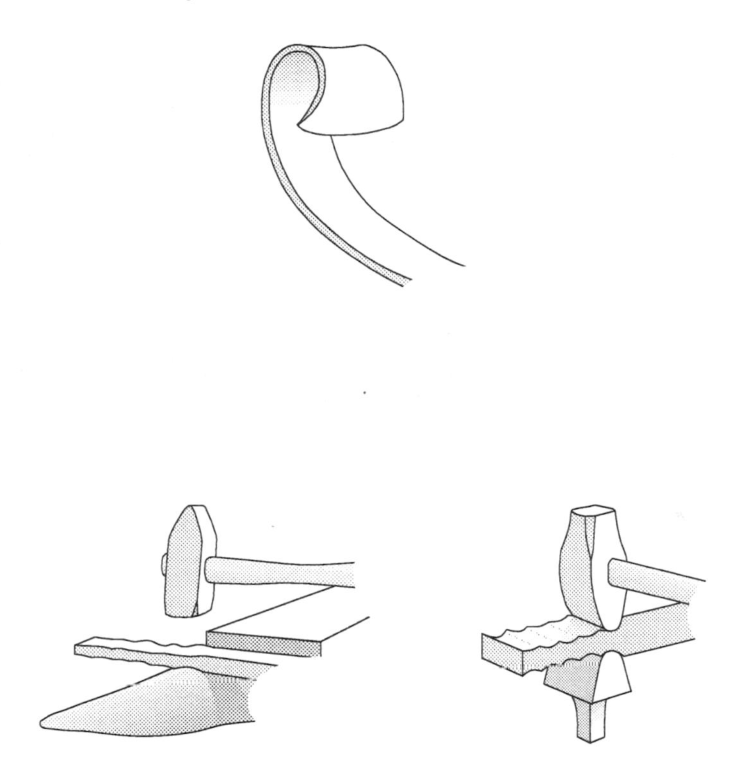

Figure 4.6. Fishtail scroll (top) and two procedures for fullering.

In Harper's work with the mechanic-engineer Willie, we find another example of constellations. Harper elicits the rationale for a constellation employed in the restoration of a field chopper, a common New England farm implement, by asking Willie to describe the activity depicted in a photograph reproduced here as Figure 4.7. The photograph depicts welding in process. Clearly evident are the welding torch, a length of metal, and a material surface being repaired. Willie explains,

On about all your farm or industrial equipment it's not just welding it together again – it's the idea of welding it so it wears right . . . In [the photograph] I'm using high-tensile steel. To tack it on I use a coat hanger. The high tensile steel doesn't

Figure 4.7. Welding constellation in action. Mechanic-engineer Willie uses high-tensile steel to make a repair. Reprinted with permission from Douglas Harper, *Working Knowledge: Skill and Community in a Small Shop* (1987), p. 71 (Chicago: University of Chicago Press). © 1987 by the University of Chicago Press. Photograph © 1987 by Douglas Harper.

have much flexibility. The coat hanger is more mild, more flexible. It's less apt to break afterwards. Vibrations break the higher tensile a lot easier. But I used it because I'm welding the bar that comes up the side – that's an important place that needs more strength. (Harper 1987:70)

We can reconstruct the implements and materials of the constellation here as welding torch, coat hanger, high-tensile steel, and the damaged field chopper. Governing this configuration of items is Willie's procedural notion of "welding it so it wears right." Wearing right is a characteristic of the desired end, for which Willie has developed a notion of means involving torch welding with a combination of a mild flexible metal and a stronger, more rigid metal to produce a weld "that is stronger than a factory weld" (Harper 1987:70). This example demonstrates that part of the conceptual component of a constellation is a concern for the physical properties and mechanical potential of the material component of the constellation.

Similarly, Schiffer's (1992) recognition of the importance of conjunction parameters in defining an activity captures much of the integrated notion of constellations. Conjunction parameters are those properties that elements must possess in order to behave appropriately in a particular activity.

> For example, in the activity of bipolar flaking of stone, the cobble hammerstone must be of adequate toughness to withstand repeated blows . . . It must also be of a size sufficient to create the momentum for fracturing the raw material . . . However, the color and precise shape of a hammerstone are not relevant for bipolar flaking. (Schiffer 1992:78)

Although Schiffer does not explicitly note the conceptual dimension relevant here, it is clear that conjunction parameters encompass the perceived affordances enabling a particular technological activity. For our purposes in developing an account of how someone makes something, the constellation represents this essential conjunction of an activity as understood by the actor and as materially facilitated by the equipment at hand.

The relations of ideas, tools, and materials in constellations

It is only through the formation of constellations for production such as those illustrated above that the open sets of possible means, ends, implements, and materials available to a smith or other actor become focused in a plan for accomplishing a goal. The constellation is a hypothesis positing a step in a procedural sequence. These arrangements of mutually constraining ideas and real-world elements are essential to the efficient enactment of a productive sequence. This is particularly clear in the case of blacksmithing because construction of a constellation allows the blacksmith to accomplish his goals in the episodic intervals necessitated by the activity of forging itself. It is the assembly of tools and materials to be at hand for enacting selected means to an end that enables the blacksmith to work efficiently within the time provided by each heat. In turn, the construction of constellations is enhanced by the rhythmic and episodic nature of blacksmithing, which provides for a period of reflection and anticipation while standing at the fire heating a segment of iron. The effectiveness of the smith's actions during the productive phase of each heat is largely a reflection of the adequacy of the constellation pulled together for the current step.

The process of originally formulating a constellation involves consideration of recognized alternatives while gradually pulling ideas, implements, and materials together. As elements are selected for the constellation, they may affect ideas or tools and materials previously selected. In addition, as elements are brought together they constrain the possibilities for further selections.

The production of any goal is problematic in the sense that alternative procedures, implements, and materials are potentially suitable for achieving desired ends (Schutz 1971). These alternatives are not coextensive with technical choices as referred to by Schiffer (1992:52–53) and Lemonnier (1993). We refer instead to alternatives known to and available to the practicing smith. We have already described alternative methods of drawing out. Choices from among available alternative procedures for drawing out will constrain the possibilities for the range of tools and materials to be included in a constellation. In addition, raw materials are chosen to achieve the goal of a task at hand and these choices have implications for the tools selected for production. For example, choosing round stock of a given size means that tongs with jaws of a certain size and shape must be used. Substituting flat stock, even of the same dimension, might require the use of a different set of tongs and perhaps a hammer with a different face. Selected tools, in turn, may have implications for notions of procedure or selected materials. Should a smith choose to use the trip-hammer for drawing out he must use a procedure for thinning and lengthening iron stock that involves the mechanical hammer. In addition, if the decision is made to draw out stock under the trip-hammer, then only tools usable with the trip-hammer can be employed. The constellation as ultimately formed is a structure of mutually constraining elements integrated (but not determined) by a logic constructed by the blacksmith from the perceived requirements of a task at hand, his general stock of knowledge regarding procedures and material properties, and his principles.

In Chapter 2 we pointed out that many of the tools used by the artist-blacksmith are unspecialized forms, any one of which can be used to produce a number of different results. The specialized components of a constellation are often the ideas formulated as hypotheses regarding the means to attain a particular end. Any given tool may be a constituent in numerous constellations, with the goal and logic of each particular step defining the use to which the implement is put. It is important to note that the ideas constituting the mental components of a constellation often include procedures for correcting or repairing deviations from the image of the desired outcome of a particular step in production. Therefore, tools may well be used in multiple ways even within a given constellation.

When discussing the stock of knowledge we demonstrated that the arrangement of the inventory in a shop reflected the general kinds of use to which tools and equipment were put and that it was not necessary for activity to be ongoing for the general logic of this arrangement to be apparent. However, such is not the case with the tangible components of a constellation. For example, a common element in architectural ironwork is a snub-end scroll. It is made by partially cutting through the end of a square bar, bending the stock back on itself at the point of the cut, and welding it. The resulting square lump is forged to a round shape and the bar then formed into a scroll. The procedures involve cutting, forging, and welding. The implements necessary are the forge, anvil, hammer, hardie, flux, and spoon. Exactly the same implements and similar procedures are used when putting a square head on a bolt. See Figure 4.8 for an illustration of both the snub-end scroll and a square-head bolt. In the case of the bolt, the end of a heated square bar is wrapped around the end of round bar. The square stock is cut off and welded in place at the end of the round stock and then forged into a square.

If an observer were to look at the equipment used after a snub-end scroll or a square-head bolt had been made, without the end product at hand it would be impossible to determine which of the two forms had been produced. If the completed work were also

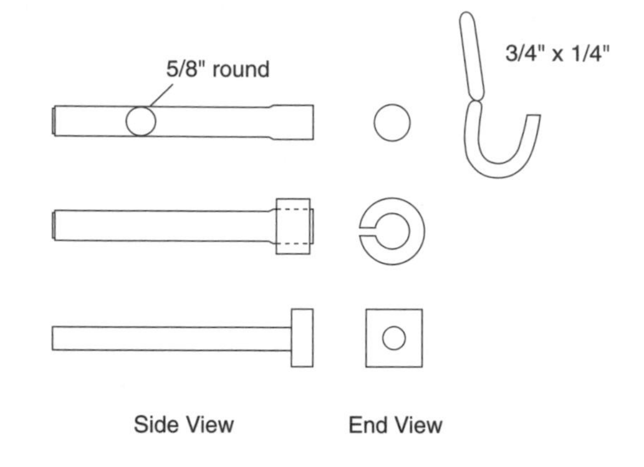

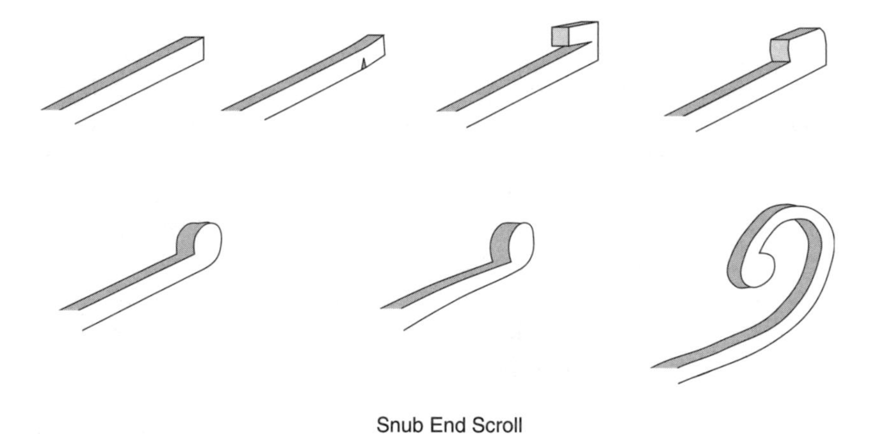

Figure 4.8. Procedures for producing a snub-end scroll (bottom) and a square-head bolt (top). The same set of implements is involved in each production, but the governing ideas are specific to the respective goals.

present it would be possible to reconstruct with some accuracy the procedures used, as blacksmiths frequently do when examining artifacts, but full understanding would be possible only if the complete sequence of procedures was observed as it occurred. The tool component of a constellation is not definitive of any specific procedure or outcome. An understanding of the orienting relationship between the task at hand and the procedures applied is possible only by observation of the situated activity.

Given the tendency of artist-blacksmiths to engage in procedures that include a novel or unaccustomed component, the open nature of constellations themselves is also important. For example, if it proves impossible to perform a particular procedure with the material elements of the constellation assembled, different or additional components must be included to complete the task. Twisting has already been mentioned several times as a common decorative element in some kinds of ironwork. During his apprenticeship CK routinely put decorative twists in pokers, shovel handles, and the elements of railings. The constellation for the procedure included a bar ½ or ⅜ inch square, an adjustable twisting wrench that would fit the stock, the vise, the straightening tools described above, and the forge. During a railing project that included a post made of a bar 1 inch square, he decided that a medial twist of one full turn would be appropriate. The requisite tools, including a wrench normally used for twisting ½-inch bars, were assembled based on a recipe derived from his past experience and his notions of procedure. The bar was heated and placed in the vise. However, a bar 1 inch square contains four times the mass of one that is ½ inch on a side, and it proved impossible to twist the larger stock with the wrench selected. No amount of straining would move the metal. Consequently, it was necessary to find a piece of pipe in the shop that would slip over the handle of the wrench and provide the additional leverage necessary to complete the twist once the bar had been reheated.

In this case, when the anticipated application proved unsuitable, recourse to prior experience was again necessary in devising a new constellation that would prove adequate to the task at hand and appropriate to the smith's general principles. Rather than abandoning the original plan for the railing project and the formulated constellation for producing a twist, CK used this information in conjunction with the problem that arose to reformulate the configuration of tools and materials needed for producing the desired twist. Perhaps it is such reformulation in a more pressing time frame that constitutes in part the recourse to embodied skill in problematic situations referred to by Suchman in her reflections on the activities of a canoeist running a series of rapids (1987:52).

Discussion and implications

The preceding analysis supports the position that knowledge is organized in a general stock reflecting past accomplishments in anticipation of future projects. This is knowledge organized for forging iron. A specific task at hand becomes conceptualized for production as an umbrella plan. The umbrella plan constitutes a conceptual framework (Keller and Lehman 1991) to be manifested in procedural steps for accomplishing a specified goal. Each step is enabled by a constellation of ideas, implements, and materials. A constellation then constitutes a hypothesis that specifies means for achieving an immediate result within the constraints of the overarching plan. Each hypothesis is subject to verification or revision in practice.

Constellations may be ephemeral, held together only as long as relevant for production, or they may endure as techniques or recipes in the stock of knowledge with tool assemblages potentially realized in the organization of the shop. The unique requirements of situated tasks create an ongoing dynamic between the stock of knowledge and constellations applied in practice. An actor has the continuing possibility of expanding his stock of knowledge by reflecting on the results of activity and refining the constellations subsequently employed in goal attainment. This dynamic relationship is the foundation of the flexibility characteristic of artist-blacksmith production and, as we will demonstrate in Chapter 7, characteristic of other activities as well.

Artist-blacksmiths are typically well aware of the potential for reproducing established practices and innovating in their practice.

In *Heartland Blacksmiths: Conversations at the Forge,* two blacksmiths discuss an aspect of this flexibility.

> Rick: . . . The flexibility that blacksmiths have is to their advantage. If the market floods in one certain area, they can shift to other areas. That's always been a strong point for blacksmiths through the five- or six-thousand-year history of blacksmithing. If the war stops and the sword business goes down, they can start on plowshares.
>
> Don: That's right. You don't have too much tied up in specialized capital equipment. You're using the same anvil that you made stove handles on, but now you're starting to make balustrades for your architectural business. You're dealing with very fundamental techniques that are applicable to a broad range of items. (Reichelt 1988:118)

It is this flexible quality of production that we begin to be able to account for through a dynamic framework that integrates mental and material aspects of a larger activity system. No one factor is consistently determined by nor independent of another. It is the essential dialectic of constellations, which, through their applications, allow the growth of new knowledge and revision of old knowledge. Continual learning is a potential consequence of such a system. In concert with the flexibility typical of the blacksmith's approach, the constellation as the enabling unit of production is open-ended, subject to the mutual influences of mental and material components in a construction of means to a desired end, and further subject to alteration and elaboration in practice.

We began this chapter by repeating the question, how does someone make something? What we have argued here is that guided by governing principles and the features of a specific task, a person forges an artifact in iron by drawing on his stock of knowledge to develop a design and conceptualize means to this end. This umbrella plan is then subdivided into procedural steps, each enabled by a constellation of procedural notions and material resources. Having assembled tools and materials, the smith then tests his constellation in action, revising it as needed. In the next chapter we continue to address the relations of knowing and doing by turning our attention to these actual sequences of productive activity and to the implications of such actions for the conceptual organization of production.

5 Emergence and accomplishment in production

> The handle on top is made of wrought iron. To get the look of old work on the straps, which are hinged, I decided to faggot weld them up from scraps. One weld gave me particular trouble, crumbling in a new spot each time I welded. I found I had mistakenly picked up a scrap of high carbon steel. Pleased with the look, I welded a mild steel bit behind the faults for strength and proceeded.
>
> (Latané 1993)

In theory

In 1957 Ward Goodenough created a new mandate for anthropologists. He directed our attention to knowledge and action by setting the anthropologist's task as that of accounting for "whatever it is one has to know or believe in order to operate in a manner acceptable to [society's] members" (Goodenough 1957:167). This often-cited passage is worth repeating because scholars in anthropology and related fields have typically focused on the first element of this directive, "whatever it is one has to know or believe," to the exclusion of the second and equally important element, "in order to operate [behave] in a manner acceptable . . ." What Goodenough's mandate requires is an account of knowledge and action in tandem: a dynamic approach to their complex interrelations.

It is toward such an account that we have directed this book. In Chapter 2 we described the principles held by the members of our subject community. In Chapter 3 we addressed the stock of knowledge as repository and as resource for anticipated activity. In Chapter 4 we examined the organization of aspects of this knowledge governing associations of tools and materials in forming constellations for action. In this chapter we take up the issue of production, or "operating" in Goodenough's original terms. This is the active side of Goodenough's mandate and is referred to here as production to avoid confusion with technical references to operations (Renfrew and Zubrow 1994; Lemonnier 1992, 1993; Leont'ev 1981). We argue that production is governed by an umbrella plan. This plan is a conceptual representation of the intended design of an object and the procedures for its production. We also explore the hypothesis that production is an emergent process involving a synergistic relationship among ideas conceptualized in the umbrella plan and the practices that enact them. We find knowledge and action are each derived from and open to alteration by the other as behavior proceeds (Suchman 1987; Holland 1992). Knowledge and action are continually in revision and development; we must understand them as mutually constructed and equally dynamic entities.

In what follows we will continue to develop an approach to the question, How does someone accomplish a task at hand? We apply ourselves to the analysis of a detailed example of blacksmithing, but the emergent relationship identified here between knowledge and action is broadly relevant to human behavior. For example, Frake (1985) provides an illustration rooted in the real-world problem solving of medieval seafarers; Hill and Plath (1996) develop another clear example in their analysis of Japanese shellfish diving. In addressing production holistically we also elaborate the account begun in the preceding chapters of whatever it is the smith needs to know to produce a specific end product.

The concept of an activity system is central to the account of production we develop. Within an activity system *actions* – strategically governed behaviors oriented toward a particular goal – are embedded in *social parameters* – frequently referred to as rules, principles, and schematic guidelines for behavior – and maintain a dialectic relationship with the available *technology* – tools and procedures for their use (see Leont'ev 1981; Wertsch 1981). The focus of this chapter is at the level of instrumentally mediated actions. Social parameters are addressed as these are represented in the form of principles and schemata and applied by a particular actor.

Our approach in this chapter can be usefully compared to that of

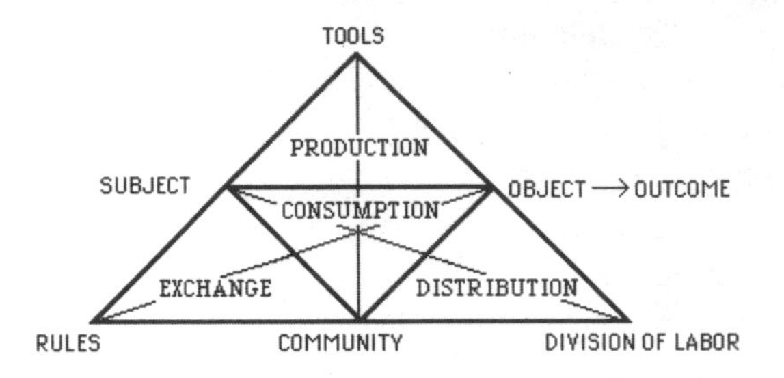

Figure 5.1. The basic structure of human activity. From Yrgö Engeström, "Developmental Studies of Work as a Testbench of Activity Theory: The Case of Primary Care Medical Practice," in Seth Chaiklin and Jean Lave (eds.), *Understanding Practice: Perspectives on Activity and Context* (1993), p. 68 (Cambridge University Press). © 1993 Cambridge University Press. Reprinted with the permission of Cambridge University Press.

Engeström (1987, 1993) in his development of a framework for the study of collective activity. Engeström places individual activity within a baseline of social phenomena, as illustrated in Figure 5.1. The relationship between subject and object here is a dynamic and dialectic one mediated by instruments and social parameters of relevance. However, Engeström focuses on the system in sociohistorical development, while it is our task to account for the accomplishments of a person acting within such a system. We, therefore, take the productive act as our starting point, drawing on instruments and social resources as these affect production, but attending neither to technoscience nor to other social parameters independent of the productive act. As Figure 5.2 shows, we differ from Engeström in our lack of attention to the division of labor and in our focus on knowledge and practice as resources rather than as rules constituting socially given parameters for production.

For our purposes herein a focus on the productive process allows us to deal with actual events in observable segments of real time. The blacksmith's activity is segmented into units referred to as heats, which represent the time during which a heated section of iron maintains a sufficient degree of plasticity to allow changes of shape by forging. These are brief periods typically lasting for only

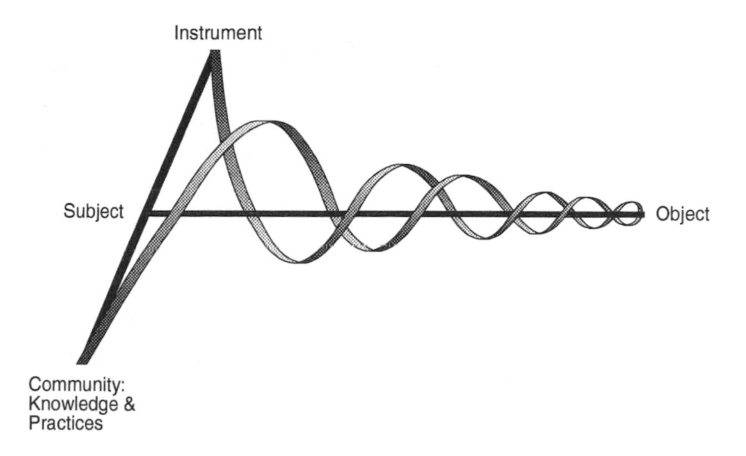

Figure 5.2. The productive act as a dynamic accomplishment.

seconds or minutes. Our analysis takes the heat as basic. We focus in detail on productive activity in a sequence of heats during which steps in production are accomplished and an artifact is produced.

In practice: design and planning

In the production we will examine, CK makes a skimmer handle, a relatively simple but not routine task for him. JK follows CK's activity, observing and directing questions to him as he works.

A skimmer is a long-handled kitchen utensil with a shallowly dished, perforated bowl, primarily used for removing residue from the top of bubbling, cooking liquids. CK had made a skimmer once before, but of a different and simpler style than the one made here. He had also produced other utensils mechanically and functionally similar to skimmers, such as ladles and spoons. Some of these production sequences had even become routine. However, none of the earlier end products conformed to the stylistic parameters required for this skimmer handle.

Our selection of a nonroutine task for study was deliberate. When a productive sequence has become routine, much of the activity entailed is "taken for granted" and is enacted without reflection. It is typically difficult for someone engaged in routine behavior to articulate the tacit premises governing their performance (Polanyi 1962; Schiffer 1992; Schutz 1971; Suchman 1987:114–115). A novel production, however, involves problematic aspects that must be resolved. This process focuses the attention of the actor on his activity, thus facilitating our analysis. Given the blacksmith's preference for undertaking nonroutine tasks, such a project was congenial and not unrepresentative of a smith's usual practice.

Initiating the conception of a task at hand for CK, the staff of a history museum asked him to provide a whitesmithed skimmer in the spirit of 18th- and 19th-century American utensils. Whitesmithed artifacts are forged in iron with the dark surface left by the forging process removed by filing. The staff was motivated in this request by a desire to replace a chrome-plated spoon attached to a stick that they had been using at occasional demonstrations of early American domestic activities. While the spoon allowed them to avoid using original museum specimens, the staff found it both functionally and historically inadequate and aesthetically inappropriate for their needs.

In presenting their request to CK, the museum staff showed him examples of period skimmers from their collection; he noted characteristics of line, proportion, and finish that could be adapted for a preliminary conception of a goal for production. As the process of developing a design for the skimmer continued, CK looked elsewhere for appropriate exemplars in addition to those from the museum's collections. He did so by drawing on his own stock of knowledge regarding the designs and procedures entailed in the production of similar implements. He also reviewed available literature depicting kitchen utensils of the appropriate period.

Other considerations relevant to the design process revolved around commercial aspects of the profession, including the time and effort appropriate for this production, particulars of the client–blacksmith relationship at issue, the material conditions for work, and financial constraints. Mental images, both visual and kinesthetic, also entered into the design process. These constitute criteria for evaluating particular design possibilities in terms of the union of function and creativity so central to the artist-blacksmith's work.

Finally, the design process was encompassed by a personal conception of the enduring principles discussed in Chapter 2. In planning a task at hand, a preference for transforming hot iron with hand-controlled tools in a production process that remains open to uncertain results governs the consideration of both specific information and general criteria.

The design process is one of moving back and forth among these diverse sets of relevant information with the task at hand in mind. In the case of period replicas, according to CK, his artifacts ought not to be exact copies of particular specimens, but rather to look as if they could have been made during the historical era in question (see also Van Esterik 1985). The request by the museum staff that the skimmer be in the spirit of 18th- and early 19th-century American utensils thus called upon CK's conception of the stylistic standards for kitchen utensils of that period: appropriate segmentations in the handle, cross-sectional contrasts, and transitions between them. It was these stylistic elements that CK considered in his examination of museum specimens and depictions in the literature and in his review of similar productions from his stock of knowledge.

The museum examples initially available were considered by CK to be extremely simple, lacking significant cross-sectional changes. He hoped to produce an artifact with a design more complex yet still within the appropriate stylistic range. At one point, while reviewing related literature, CK expressed his "desire to produce something more elaborate than the examples in the museum yet still in keeping with the period." Figures 5.3 and 5.4 illustrate the kind of published material to which CK turned to familiarize himself with the period constraints on style so that he could use them to generate a novel but historically fitting goal for production.

Mental images of line, proportion, contrast, elegance, and simplicity were also brought to bear in evaluating the stylistic possibilities appropriate to the historical period at issue. The aesthetic and stylistic criteria taken together suggested an open range of possibilities that would satisfy the requirement that the work must look good and conform to the standards of the historical period.

Criteria relevant to mechanical adequacy also entered into the process here. (For a similar discussion in terms of performance

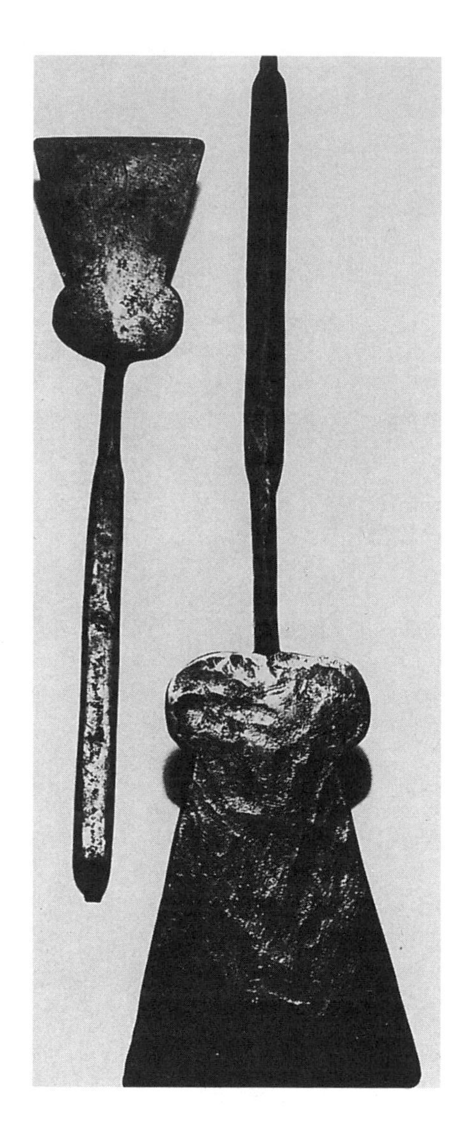

Figure 5.3. Kitchen utensils. Reprinted with permission from Jeannette Lasansky, *To Draw, Upset and Weld: The Work of the Rural Pennsylvania Blacksmith 1742–1935* (1980), p. 51 (Lewisburg, PA: Oral Traditions Project of the Union County Historical Society). © 1980 Jeannette Lasansky and the Oral Traditions Project.

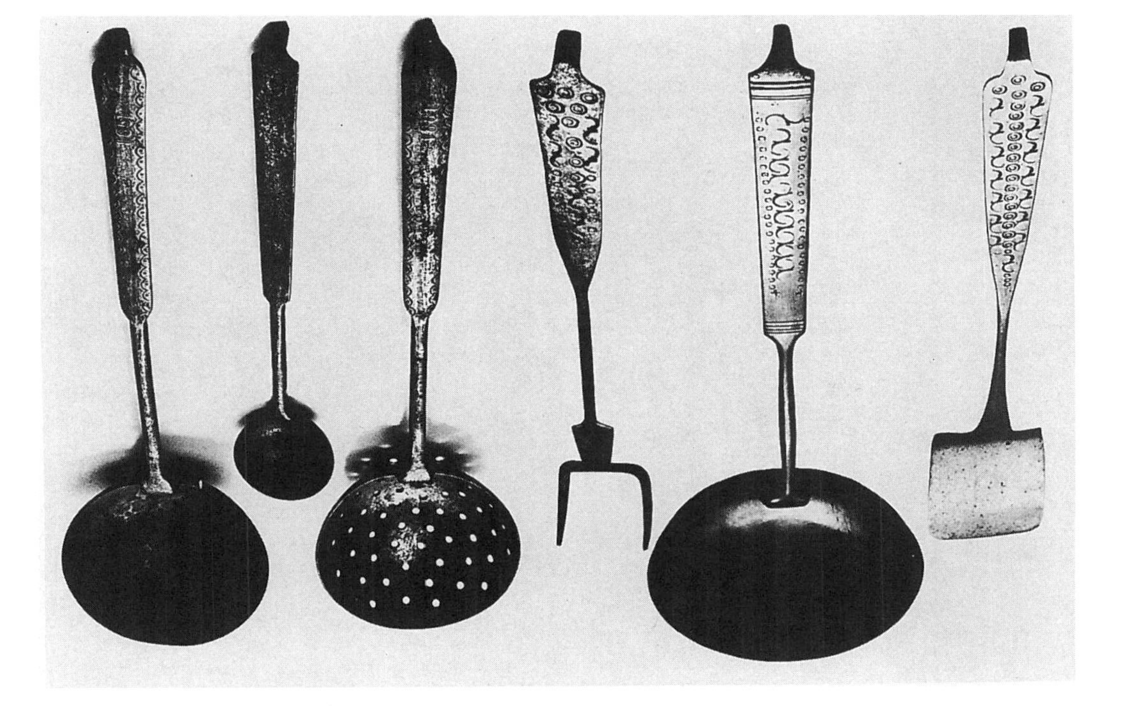

Figure 5.4. Kitchen utensils. Reprinted with permission from Jeannette Lasansky, *To Draw, Upset and Weld: The Work of the Rural Pennsylvania Blacksmith 1742–1935* (1980), p. 71 (Lewisburg, PA: Oral Traditions Project of the Union County Historical Society). © 1980 Jeannette Lasansky and the Oral Traditions Project.

characteristics, see Schiffer 1992.) CK did not consider the planned skimmer purely as a sculpture but aimed to produce an item that worked. The handle had to be long enough to prevent injury to the cook using it. It had to be balanced with the bowl to facilitate use and be comfortable to grip. Such requirements provided another set of features that would characterize the final product yet again leave a wide range of acceptable possibilities open for the actual design.

Functional and aesthetic criteria in turn called up questions of procedures for accomplishment that were further relevant to the design process. How does one produce possible design features in iron, and what techniques are appropriate for the reproduction of early American products? Rinedollar, an artist-blacksmith who frequently works in traditional styles, puts it this way: "If you want to make strictly traditional stuff, then you use the traditional tools. If you want to compromise a bit, you use a little of both [traditional and modern technology]" (Rinedollar in Reichelt 1988:54). As usual, the blacksmith must choose among alternatives.

In designing the skimmer CK considered questions of the technical requirements for accomplishing particular aesthetic and functional aspects of the product. He recalled previous productions relevant to the present task, particularly period kitchenware. These recollections suggested possibilities for production given by prior constellations and procedures that were either successful or problematic. CK also considered his own skill in carrying out particular techniques and the material conditions for his work, including the shop, available tools (always with the possibility of producing or purchasing new ones), and raw materials. These considerations situate the aesthetic and functional possibilities for design within considerations of the material and procedural possibilities for production.

Criteria relevant to the commercial aspects of blacksmithing were also considered. As Rinedollar indicates, it is usually important to evaluate the possibilities for production in terms of the economic constraints on the smith. CK considered how much time and effort to spend on the skimmer in order to "come out even" (Rinedollar in Reichelt 1988:52). This involved considerations of income and costs of production balanced against the pleasures of the work process itself, the functional and aesthetic possibilities for the project, and

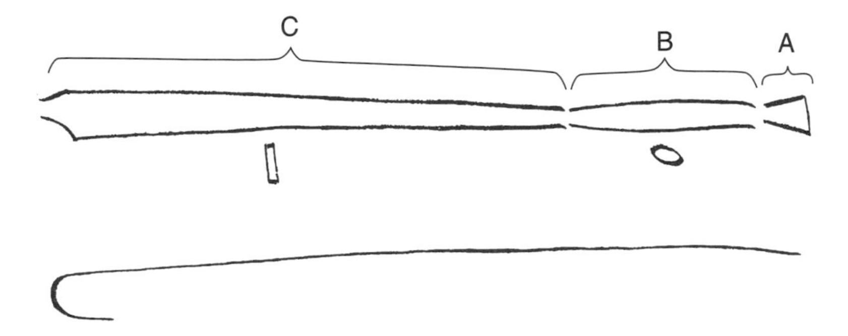

Figure 5.5. Design sketch of 19th-century-style skimmer handle, drawn by CK prior to initiating production.

the value placed on working with traditional techniques. Involved here as well was a consideration of CK's self-image as academic and smith and his conception of the expectations of his client. These values, standards, and criteria operated as influences on design rather than determinants of outcome.

In practice: graphic representation

Ultimately CK decided on a particular design for the handle alone, leaving the bowl for later. It is in the production of this handle that we examine the relations of knowledge and action. The design arrived at constituted an initial goal orientation for the task at hand. CK outlined this goal in the sketch presented in Figure 5.5.

A number of things about this figure are worthy of note. CK has sketched the anticipated end product in vague outline only. Details of the completed skimmer handle are not evident. This was in part the result of using photographed examples in the literature in developing design possibilities. When the design process depends on reference to two-dimensional depictions, the smith is left to construct his product in three dimensions from mental images. In this case CK's sense of specific design options was less detailed than examination of a greater range of material exemplars might have allowed.

In addition, the sketch is a graphic depiction that lacks reference to the anticipated production procedures. A rough idea of the procedural sequence was implied for CK by the sketch: draw out and shape bearing surface A; draw out and taper segment B; draw out and flare section C; narrow the stock to form a linear extension to be bent into a hook; and finally, make the transitions. These procedures, selected from among possibilities given by his stock of knowledge, were developed by CK as he formulated the design plan. However, the vagueness in the design conception itself precluded development of a complete procedural sequence at the outset. For instance, no procedures for making the transitions were specified, since no conception of these features was present in the original design outline.

This vagueness and openness to development is an inherent characteristic of goal orientations for the productive activities of blacksmiths. The same quality characterizes the plans or mental models of inventors (Gorman and Carlson 1990:136) and practitioners of crafts other than blacksmithing (Gladwin 1970; Hutchins 1983; Hill and Plath 1996) and industrial designers (Vincenti 1990). Even the most thorough design and associated plan for action can never be sufficiently detailed, sufficiently precise to anticipate everything that can happen during production (Suchman 1987:185–187). The process of production always has the potential to force changes in plans.

Artist-blacksmiths rely on emergent characteristics that result during production to complete even the most detailed design. As one blacksmith says, "You have to follow the material" (Reichelt 1988:50). An initial design is an orientation toward production of an artifact that the blacksmith anticipates will acquire an increasingly specific character as work proceeds. In this vein, the well-known blacksmith Samuel Yellin once commented to a group of architects, "There is only one way to make good decorative ironwork and that is with the hammer at the anvil, for in the heat of creation and under the spell of the hammer, the whole conception of a composition is often transformed" (Andrews 1992:72). Making something in this way is a process of gradually specifying the product both mentally and materially in the very act of producing it.

In the end – or perhaps more aptly at the beginning of production – CK articulated a decision "to make a skimmer handle like the leftmost example [in Figure 5.3], but with more obvious changes in cross-section and in transition between the major thirds of the handle" labeled A, B, and C on Figure 5.5.

This combination of design and implied procedure we have referred to earlier as an umbrella plan, a mental representation of the overarching goal and a rough procedural sequence for its attainment. Notice that the umbrella plan, while developed for a specific task and open to further specification in practice, was also consistent with the principles of transformation, working hot, and freehand tool use. In line with the basic premises of his practice, the smith crystallized an idea for the goal for production as depicted by Figure 5.5 and developed a preliminary procedural orientation sufficient to start work.

This umbrella plan was a proposal in the form of a schema for goal attainment. Note that the schema was not simply an empirical generalization from exemplars (Gopnick and Wellman 1994), although this was a significant process in its formation. Also entailed was the crystallization of a procedural plan. This required selective review of the stock of knowledge representing procedures, past productive experiences, and anticipated future possibilities. This knowledge was highlighted and reviewed against the dimensions of relevance for this task. The umbrella plan proposed here was again like the mental models that Gorman and Carlson (1990:136) attribute to inventors as they imagine a new technology, and craft and manipulate prototypes. It was a plan for action crucially containing the expectation that further material and procedural details beyond those specified would emerge in the process of forging the artifact.

With the umbrella plan completed, CK was now on the verge of engaging directly in productive activity. As we will see below this plan would be detailed, enriched, and altered as CK's internal representation confronted the objective conditions of work. Details would emerge in the production sequence, perhaps replicating previous solutions to production problems, perhaps creating novel alternatives. We now turn to an examination of this productive activity itself to explore the relations of knowledge and practice.

In practice: accomplishing a goal

In this section we detail the particulars of the productive activity involved in making the skimmer handle. We focus on actions divisible into the following steps: heating, the goal-oriented transformation of the material, evaluation of the result, and transition to the next action (Zinchenko and Gordon 1981:93). We do this with two objectives in mind. First, we hope to demonstrate the relation of the internally represented umbrella plan to the structure of action. Second, we hope to demonstrate the effects of ongoing actions and their material results on the organization and substance of the umbrella plan and constellations that develop within it. Within this second objective we also consider the implications of the accomplishment of production for the stock of knowledge.

We will use Figure 5.5 for reference to the handle segments throughout the following discussion. Segment A will ultimately be the bearing surface for the skimmer bowl. Segment B is to be a rounded midsection tapered at each end. Segment C will be a longer, flared flat section that culminates in a transition to a hook for hanging the skimmer. While these segments are evident in the sketch, most of the detail, particularly of proportions and transitions between segments, was left to be resolved in practice.

The umbrella plan directed the initial stages of action. CK selected the stock for the skimmer handle by evaluating the possibilities of particular raw materials against the requirements of the goal. In his words:

> Since it is easier to make round stock flat than it is to make flat stock round, a piece of round stock is chosen. The diameter of the stock is approximately the same as the maximum diameter of the midsection of the handle to keep forging to a minimum, but it must have enough mass to be spread in both end sections without getting too thin [as judged both functionally and aesthetically].

With his stock selected and cut to an appropriate length for the anticipated skimmer handle, CK prepared the fire. He gathered tools around the anvil in anticipation of the early procedural steps, which would involve drawing out the stock. The tools of this early constellation included a three-pound, flat-faced, cross-peen hammer and a pair of tongs suited to the dimensions of the stock. He lubricated the trip-hammer, which he had also decided to use in drawing out and shaping segment A. The trip-hammer was incorporated in the initial constellation to allow the early stages of drawing out to proceed more quickly than would have been possible exclusively with hand hammers. The final stages of drawing out and shaping would be done with the hand hammer. These preparations reflect the anticipation of actions to follow.

To this point the relationship between the umbrella plan and action is largely one of internal direction. But the process does not continue in a one-way fashion. As we examine the steps in the production of the skimmer handle, the significance of material results for directing the continuing activity become clear. The overarching goal of production was the skimmer handle, but each step of the way was oriented to more particular goals, such as flattening and shaping sections A and C, tapering the round middle section by drawing it out toward each end, and forging the hook at the end of section C. Each of these productive sequences influenced the umbrella plan and affected subsequent actions.

CK began with section A, using both the trip-hammer and hand hammer to flatten the bearing surface. This action, seemingly straightforward at the outset with respect to both the immediate goal desired and the procedures to be followed, allowed CK to anticipate upcoming problems as he worked the metal. We say this action was seemingly straightforward here because, to get ahead of ourselves just a bit, even bearing surface A was reforged several times at later points in the productive sequence after evaluations of the results of subsequent actions. In any case, even in these early stages, the implications of the lack of specificity in the image associated with the umbrella plan were apparent. While CK was working to flatten the bearing surface of the skimmer handle he was thinking not only about the shape segment A should take but also about the transitions between segments of the handle with different cross-sections.

As our discussion of the production of the skimmer handle proceeds, we will pay particular attention to the development of these transitions. The emergence of their material form illustrates well the potentially directive role of activity and its results in production. However, the process by which the transitions took shape was repeated, though perhaps less dramatically, at each step of the activity.

CK decided initially to "take care of the transitions when filing." Filing is a cold, subtractive technique that could appropriately be employed here in the final stage of production. This initial decision left the most problematic details to be resolved late in the productive sequence.

As CK continued to work to flatten segment A and then to taper section B, he reviewed the decision to file the transitions. As he reflected, he considered that he wanted "to file as little as possible, so . . . [I] must forge cleanly" to avoid the extra filing work that would be needed to remove unwanted irregularities. To the tool assemblage of the constellation for drawing out the iron, he added a wire brush to remove the scale after each heat and so keep the surface of the work as clean as possible. He then commented that "forging is more interesting than filing so . . . [I] will forge as much as possible." Notice that forging more and filing less is consistent with the principle of thinking hot. Following this reconsideration and in contrast with his earlier decision, CK then decided that he would forge the transitions between segments.

Clearly, while CK was initiating production he was continuing to review his umbrella plan in light of his fundamental principles for work. In recognizing the choice to file as divergent, he reassessed his plan and altered his procedural expectations for upcoming steps. His reflections at that moment were not restricted to the step at issue but took him conceptually back over the full umbrella plan.

At this point he attempted a transition between segments A and B that failed aesthetically. He corrected an offending bulge at the transition point and then left this transition unfinished, continuing work on the rounded midsection while still contemplating the transitions.

While forging the handle segments, CK examined the work often to get a sense of proportion, contrast, possible transitions, and straightness. A number of problematic issues needed to be resolved as production proceeded. CK was trying to determine, for example, "how long [aesthetically, functionally] the tapered midsection should be. The two flat parts [A and C] have to be about the same width and have enough contrast between them and the middle to show something's happening [in the change in cross-section]." CK went on to articulate another aspect of the problem of proportion relevant here. "I can't make the round part too long because then its diameter would be too small at the transition to the flat parts. That would make it weak and make the contrast with the flat section too striking." These two features, relative length of segment B and diameter at the transition, had to be balanced to produce a subtle but noticeable contrast. Although the smith was aware that balance would be essential, the particular material balance did not come directly from the umbrella plan, but emerged as CK responded to and evaluated the objective results of his actions in drawing out segment B.

By constantly monitoring and evaluating the material results of drawing out and tapering segment B, CK achieved a satisfactory formal and functional balance between segments A and B. He then moved on to segment C.

> I'm close enough to where I want to be to start spreading the upper part [C]. There are still a few bad spots I'll have to get rid of [in section B] and I'm still not sure of the transition. I'll start spreading just above the round section [B] being careful not to spread too much at this point because otherwise there won't be enough mass at the top end [to continue the line appropriately]. I want to spread minimally at first to avoid having to reduce, because I can always spread more but it's a hassle to reduce.

During the next several heats CK spread section C again, using a combination of trip-hammer and hand hammer. In this action sequence, although he cautioned himself against spreading too far, CK did just that. It was as though he had to cross a material boundary to discover the appropriate limits of the design. Again the form emerged not from simply matching the object to the imagined goal but rather from CK's continuous evaluation of the objective results of his efforts. The idea was not sufficiently definite prior to enactment to serve as an exact standard or template. The outcome had to be assessed and reassessed as it took material form. The same values

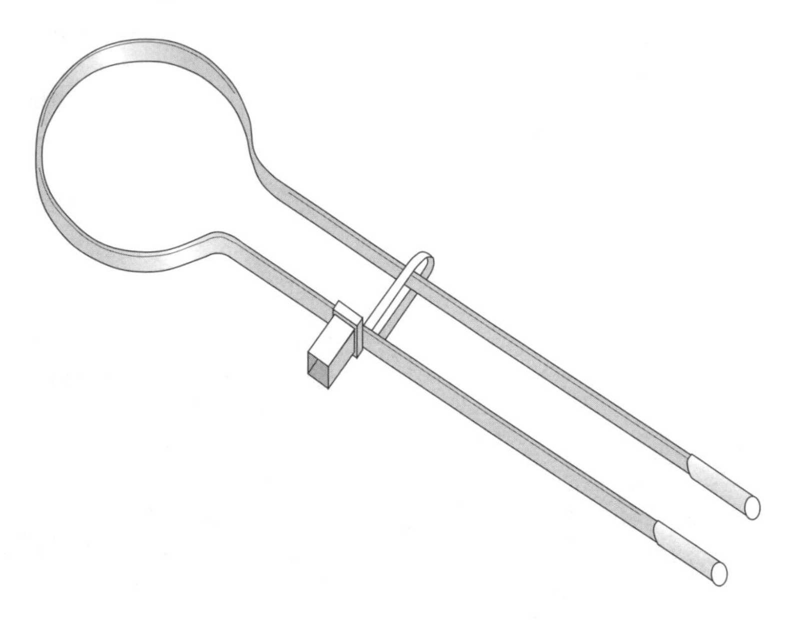

Figure 5.6. Spring fuller.

and criteria used in the original design process were employed in constantly monitoring the artifact in production. The umbrella plan was detailed and elaborated as the blacksmith reflected upon the results of his efforts; the newly detailed plan then facilitated further action.

Having spread and flared section C, CK then pulled out a tool not among those previously assembled for the constellation for drawing out the segments of the skimmer handle. The tool was a spring fuller that CK decided early in the work he would use in narrowing the now-flared end of section C at that point where the hook would begin. The spring fuller, illustrated in Figure 5.6, has the shape of an elongated U resting on its side. Stock can be inserted between the arms of the tool. Hammer blows applied to the upper arm of the U then serve to compress the spring fuller around the work, squeezing and thinning the stock.

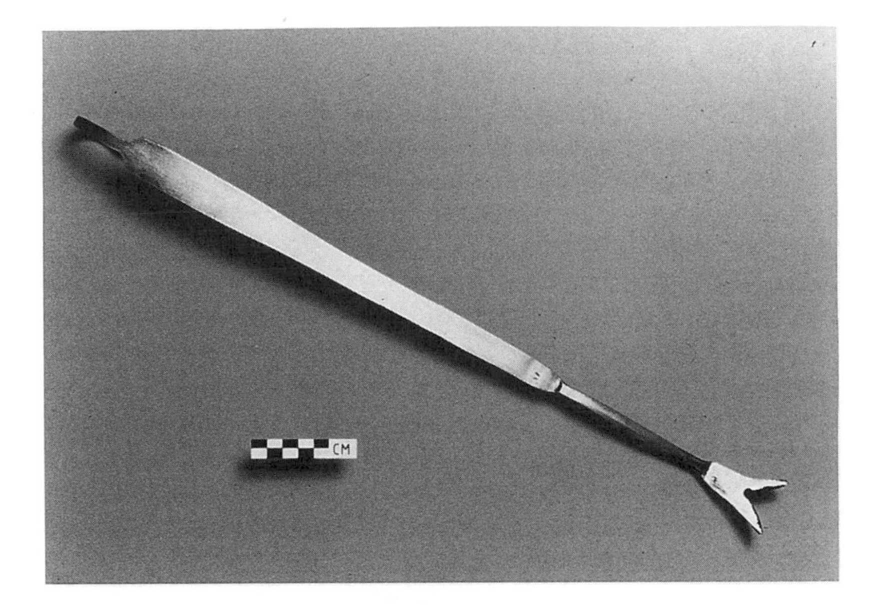

Figure 5.7. Skimmer handle. Charles Keller, 1987.

It was while examining the result of this fullering that CK finally conceived of the appropriate design for the transitions between the handle segments. Upon narrowing the flared end of segment C where the hook would begin, CK recognized the possibility of developing transitions similar in form at each segment change.

In response to the current shape of the iron and with new insight, CK decided to make the C–B transition an inversion of the shouldered and narrowed form of the material at the place where segment C makes a transition to the hook. Where the C-hook transition creates a concavity, the C–B transition is convex. The A–B transition was then shaped in the general style of the others. Figure 5.7 illustrates the completed handle.

This outcome resulted from a synthesis of the material conditions and conceptual dimensions of the project (see Shanker 1991 for a useful discussion of insight). The process was a dialectic that again

Figure 5.8. Completed skimmer. Charles Keller, 1987.

illustrated the account of invention developed in the work of Gorman and Carlson, who argue that "an inventor combines abstract ideas with physical objects or what we call *mental models* with *mechanical representations*" (1990:133; emphasis in the original).

The transitions finally accomplished, CK was finished forging the handle. Yet even after he had completed forging the transitions and after he had gone on to the final whitesmithing step, CK continued to evaluate the skimmer handle against the now-detailed conceptual representation of it and against standards of aesthetics, style, and functional adequacy. He returned at least twice to reforge segments of the handle, ultimately changing the shape and thickness of the bearing surface to provide a larger area for attachment of the skimmer bowl, and thinning section C as it approached section B to accentuate the cross-sectional contrast between them. Figure 5.8 shows the completed skimmer.

The production of this skimmer will form part of CK's stock of knowledge, perhaps as an exemplar, perhaps as a recipe, perhaps as both, and will be reviewed as relevant for future productions. The potentially integrative role of transitions, and particularly of inversion, may be abstracted from the experience and applied as a higher-order generalization to future tasks. In this way the stock of knowledge is continually expanded and reorganized to facilitate the accomplishment of later projects, both in terms of the specifics of previous projects and the general relevance of insights thereby derived.

In conclusion

We argue here that someone accomplishes something by conceptualizing a task at hand as a more or less open-ended goal (Schlanger 1990). An umbrella plan that integrates the overarching formal goal with an outline of a rough procedural sequence for attainment is created before work is begun. Enacting the initial steps and evaluating the material results of that action allow an individual to revise and elaborate the umbrella plan and further specify steps for attaining the increasingly detailed representation of the end product. The sequence of procedures is subject to reconceptualization during enactment, as illustrated by CK's early attempt at forging a transition. Ultimately, accomplishment of a project feeds back into the stock of knowledge. This "marriage of the hand and the mind in solving practical problems" constitutes what Harper calls the unity of work (Harper 1987:118). It is the unity of work that provides for the emergent essence of human productivity and for the characteristic fulfillment experienced by artist-blacksmiths in their practice.

The productive activity we have described is essentially nonlinear. Reflection involves review of higher-order relations as planfully conceived rather than simply sequentially linked steps (Kline 1984). Progress from one step to another is predicated not simply on completion of preceding steps, but on the continual evaluation of ongoing activity and its material results in light of the increasingly detailed umbrella plan. Each action undertaken is initiated on the basis of a significant fit with the current holistic understanding of the product to be constructed and the procedures for this construction.

The process is one in which internal representations and external actions and objects are dialectically integrated. A tension exists between relevant knowledge as constructed for production, and the unfolding experience (see Lutz 1992:186). This dynamic of situated action and represented knowledge gives an essentially emergent character to production. As Gorman and Carlson (1990:154) have also shown, this is the crux of invention, wherein "experiments with actual mechanisms lead to revisions in mental models." An individual's knowledge is simultaneously prepatterned as the stock of knowledge and aimed at coming to terms with ongoing actions and products that go beyond those preconceived (Dougherty and Fernandez 1982:823). Action, in turn, has an emergent quality that results from the continual feedback from external events to internal representations and from these revised notions of design and procedure back to enactment.

In addition, the creation of novel products and practices involves the ability to work within dimensions of relevance given by a task and within the constraints of enduring first principles, such as those for transformation, thinking hot, and working freehand. But it involves as well the ability to see new relations independent of preconceived dimensions as a result of the ever-present affordances that may transcend routine or typical patterns or applications.

We return to the question, What is it that the artist-blacksmith needs to know to produce artifacts in iron? One needs to know enough to conceptualize a socially acceptable orientation toward a goal and to provide a combinatorial arrangement of previous knowledge in the service of new, and therefore partially unknown, production. Such knowledge includes principles relevant to the activity of interest, general images for evaluation of aesthetics and criteria of the mechanical adequacy of a product, specifics of the relevant style, characteristics of exemplars, and economic and personal constraints and penchants. This information is not a set of rules as conceptualized in Engeström's diagram (Figure 5.1), but rather a resource pool for the evaluation of designs, procedures, and material results for a particular project (Suchman 1987).

Beyond the point at which an umbrella plan is first conceived, what one needs to know to produce an artifact becomes, in part, a product of productive activity directed toward that end. The very practices that emerge from the resources of prior structures have the potential to reproduce or transform those structures (Ortner 1989:12). This is a synchronic process but one very like the "structure of the conjuncture" that Sahlins (1985) has developed fruitfully in his account of culture and history.

The umbrella plan provides an orientation for action (Suchman and Trigg 1993). As constellations are formed and the procedural steps are initiated, any one or more of the features of the umbrella plan may become problematic. At such a point the internal representation is found wanting and may be enriched or altered as it is reconstructed on the basis of the results of action. "Whatever one needs to know" cannot be definitely specified at the outset of activity but remains open to insight and to new information that may emerge as one acts on the basis of what one already knows.

As we conclude this chapter it is important to reemphasize the diversity of cognitive modes involved in the conceptual processes we have been discussing. The reasoning, planning, evaluation, and insight at issue here are not primarily linear nor primarily verbal. Instead, causal, relational, and synthetic reasoning rooted in visual imagery and kinesthetic representations predominate in the conceptual side of making something. Thus we turn to a closer look at imagery and production in our next chapter.

6 Imagery in ironwork

> Our thoughts most often come to us in the guise of linguistic imagery.
>
> (Jackendoff 1992:11)

Language has held a privileged place in the research of scholars interested in the workings of the mind, so much so that verbal reasoning is often considered the only modality of conceptual thought. In this chapter we offer a demonstration of the crucial role of visual, kinesthetic, and aural imagery in reasoning. We will argue that nonverbal reasoning is essential in the process of establishing a goal orientation, monitoring productive activity, solving problems, and achieving a desired end. In brief, our data suggest thoughts may actually occur to people most often in nonlinguistic form, especially when they are involved in making something. We will argue that this nonverbal component of mental activity constitutes an essential dimension of conceptual thought. Conceptualization, then, is the characteristically human capacity to integrate diverse forms of information in active reasoning.

In the preceding chapters we have characterized productive activity as an emergent, goal-oriented process in which ideas and material results continually inform one another in dialectic fashion as a blacksmith makes something. Using his stock of knowledge as a resource, the blacksmith develops an umbrella plan comprising a skeletal image of the goal of production with an associated procedural outline. Against the background of this plan, the blacksmith assembles a constellation consisting of his notions of anticipated means and the associated implements and materials needed for each procedural step. This constellation enables the enactment of production, during which the results of forging may further influence the stock of knowledge, the original plan, and the current or upcoming constellations. Nonverbal imagery is essential for successful enactment of each of these aspects of producing an artifact in iron.

Previous treatments of imagery and visualization

The importance of mental images for production and more generally for cognition has often been noted (Bunzel 1929; D'Andrade 1991; Gorman and Carlson 1990; Cochran 1981; Cooper 1991a, 1991b; Ferguson 1977; Gatewood 1985; Hindle 1981; Hindle and Lubar 1986; Hutchins 1983; Johnson 1987; Keesing 1993; Kolers and Smyth 1979; Kosslyn 1981; Lakoff 1987; O'Neale 1932; Pinker and Bloom 1990; Pye 1978; Wallace 1978; Whitten and Whitten 1993; Wynn 1989). These scholars have focused primarily on visual images and have selectively emphasized structural or functional properties of such imagery.

Functional approaches have addressed the role of nonverbal imagery in production. For example, Bunzel (1929) refers to the fact that her Pueblo informants mentioned dreaming of designs and motifs to be used in pottery decoration. More recently, Whitten and Whitten (1993) provide similar evidence for visual imagery as a resource for design symbolism. O'Neale (1932) points out the mental gymnastics required to balance design elements in twined basketry artifacts. She is referring here to the visualization and anticipatory rotation of images required for planning a design that will emerge on the surface of a twined object.

Turning to production in industrial contexts, Anthony F. C. Wallace argues in his ethnographic study *Rockdale* that visual reasoning is critical to accomplishment. "The machinist thought with his hands and eyes and when he wished to learn to communicate he made a drawing or a model" (Wallace 1978:212). He goes on to claim that "the thinking of the mechanician in designing, building and repairing tools and machinery had to be primarily visual and tactile" (1978:237).

Ferguson, a noted historian of technology, echoes these comments in his discussion of the thought processes involved in engineering design:

> Many features and qualities of the objects that a technologist thinks about cannot be reduced to unambiguous verbal descriptions; they are dealt with in his mind by a visual, nonverbal process . . . (T)he designer . . . thinks with pictures. (1977:827–828)

Hindle and Lubar, historians of technology, state that

> designing a machine requires good visual or spatial thinking. It requires mental arrangement, rearrangement and manipulation of projected components and devices. It usually requires a trial construction of the machine or at least a model of it and then more mental manipulation of possible changes in order to bring it to an effective working condition. (1986:75)

And finally, Gorman and Carlson (1990), a psychologist and a historian working together, illustrate in detail the dialectic between mental configuration and model building as hypothesis testing in their studies of invention.

Studies such as these point us clearly to the significance of imagery in the production of material goods. Other scholars have investigated the structural properties of visual images (Kosslyn 1980, 1987; Marr 1982). A major concern in this scholarship has been the development of a computational model of the mind's capacity to construct visual representations. Researchers have broken the subjective experience of visualization down into component systems, elements, and processes, such as the capacity to construct two-dimensional, two-and-a-half-dimensional, and three-dimensional representations, and the abilities to recognize an object from multiple vantage points, to separate images into component parts, and to mentally rotate visual representations. Anatomically and functionally distinct but interacting systems for determining relative spatial position and identifying three-dimensional objects have also been recognized (Jackendoff 1992). Through experimental research the properties of the visual system have been outlined, uniquely distinguishing this cognitive mode from other modular capacities. Pinker and Bloom (1990) note the selective adequacy of distinct human cognitive systems for representing different kinds of information. In particular, they point out the advantages of visual representation over language-based propositions for encoding relations of Euclidean geometry, topology, and fine-grained distinctions in relative location.

Another group of scholars has explored the role of visual imagery in language-based concept structuring. Here the emphasis is typically on skeletal or schematic visual forms. For example, Johnson (1987) and Leslie (1994) identify universal qualities represented in the mind, such as verticality, containment, and force dynamics, which they argue are visual schemata constituting a foundation for conceptual thought. Lakoff (1987) recognizes the significance of image construction in the repertoire of human cognitive capacities involved in categorization. Lakoff (1987), Johnson (1987), and Keesing (1993), among others, extend these ideas to account for the foundations of metaphor in visual depiction.

The nature of imagery

The body of research just discussed constitutes a complex yet currently unintegrated set of information on visual thought. We propose to build on this work to develop a perspective encompassing both functional and structural properties of imagery in conceptualization. Our data suggest that visual and other forms of nonverbal imagery are the primary components of the conceptual structures of blacksmithing. Here we include the first principles – transformation, thinking hot, and working freehand – and more particular images of practices, artifacts, and tools. The principles, which may have rich visual components themselves, constitute a coherent framework for the acquisition, manipulation, and construction of the more specific schemata.

We want to make clear the distinction between a visual image, as we use the phrase, and the idea of a mental template as the concept is used in the literature of archaeology and material culture (Deetz 1967). A template is defined as "a pattern or guide . . . used in manufacturing," "a usu. thin metal pattern," and "a chart" (*Webster's Third New International Dictionary*). Thus the use of the term as illustrated by the following quotation – "The production of most artefacts, for instance of such stone tools as Acheulean handaxes, involves the use of a mental template, which serves to guide the craftsperson producing the artefact" (Renfrew 1994) – suggests a mechanical and repetitive application of a more or less two-dimensional pattern by an actor. By contrast, *visual image,* as we use the phrase, may refer to a mental representation of a quality, an object, or a process. We intend something that may be mentally manipulated, rotated, rearranged, revised, or carried out. While some images are relatively simple, such as those of color or straightness or fairness, others are complex, multidimensional and multimodal entities that are not simply mental jigs but rather internal rehearsals of anticipated operations or forms.

It is also important to distinguish imagery in general from graphic depictions. When we refer to visual images or visual representations we are referring to mental constructs in the mind's eye. By contrast, graphic depictions refer to actual drawings or diagrams.

The inventory of information in the blacksmith's stock of knowledge is largely composed of imaged schemata. As the focus of substantive knowledge, such schemata constitute a frequent topic of conversation among blacksmiths, thus providing extensive opportunities for investigation. The time has come for rich ethnographic documentation of such imagery in human thought.

Specifically, we hypothesize that nonverbal imagery is a significant component of both declarative and procedural knowledge. Images represent both forms and processes and exemplify the structural and functional properties outlined above. We came to recognize the importance of imagery for production in our earliest attempts to model the knowledge required for blacksmithing (Dougherty and Keller 1982). CK's reflections on his activities as a blacksmith and observations of and interviews with other craftsmen indicated that we would be unable to begin to account for what it is the blacksmith knows without providing an account of the knowledge represented in nonverbal imagery.

There is perhaps no activity in which imagery is more critical for appropriate production than blacksmithing, although as the works cited above indicate, imagery is prominent among peoples involved in the full array of preindustrial and industrial activities. As Bealer writes in his volume on *The Art of Blacksmithing* (1976:2), "The powers of visualization are far more important than brawn to the smith." In a passage quoted earlier (see Chapter 2), Richardson says of the smith, "If he can create beautiful forms in his mind, and with his hands shape the metal to those forms, then he can see poetry in his work" (Richardson 1978). We devote this chapter to articulating a framework for understanding the nature and role of imagery in productive activity.

The sensorimotor and aural components of imagery

Many of the examples of imagery cited in the research summarized above are visual. However, sensorimotor and aural imagery are often equally important in the conceptualization of productive activities. These additional modes of information processing may yield images complementing visual schemata or, more often, may be integrated with visual information in the multimodal schemata constructed in planning and production. For example, the act of recognizing a critical juncture in an operation or the act of anticipating a sequence of movements may employ a suite of images channeled through different modalities. Qualities of weight, sound, resistance, and balance are the kinesthetic or aural counterparts of form or color as procedures are chosen and work unfolds (see also Gordon and Malone 1994:39).

For instance, as metal cools during forging its color changes, but at the same time the reaction of the iron to the hammer and the way it feels and sounds when struck change as well. The soft thud of a hammer on hot iron is replaced by a romanticized but annoying ringing sound that indicates as clearly as color that it is time to reheat the work. And when hot-punching it is important for the smith to stop striking the punch when he feels that it is no longer compressing the iron. At this point he must turn the piece over, drive the punch through from the other side until he feels it hit bottom again, and then place the work over a bolster or the pritchel hole and drive out the remaining slug of metal. If he fails to feel the bottom when working from either face he might upset the end of the punch in the hole, making it difficult to withdraw.

Describing the process of forging a knife blade, one cutler says, "I have to keep in mind . . . [whether] it is a heavy knife or a light knife." The images he uses for reference for such qualities during the forging procedures are both visual and tactile. One expects visual and tactile images to be congruent. This expectation is violated when one picks up an object that looks solid, but proves to be hollow and jumps off the table; or when an object one expects from its appearance to be balanced feels unequally weighted. For the smith in these cases, sensorimotor imagery has to do with how it feels to hold a heavy or light blade and how a balanced and symmetrical blade feels, as well as how these qualities look. The judgments made while forging are based on these and other complex sensorimotor as well as visual images (Keller and Keller 1996).

Gatewood has argued in discussing the knowledge and practices of fishermen, "One experiences visual imagery and muscular tensions appropriate to certain actions" (1985:206). These forms of mental representation are an integral part of the knowledge that governs acceptable practice and can be acquired only through practice. Observation and interview are inadequate methods for the construction of such knowledge. Information must be egocentrically represented in the sense that the images rely on bodily experience and frequently take the actor's perspective. Such knowledge arises only in the process of work oriented to the production of a specific goal and must be acquired through engagement in that process.

Harper maintains that "there is a kinesthetic correctness to Willie's method" (1987:117), what Pirsig (1974) has called the "mechanic's feel." Harper continues:

> Willie's working method builds on a detailed knowledge of materials and develops precisely the kind of tactile, empirical connection that leads to smoothly working rhythms, appropriate power and torque, and the interpretation of sounds and subtle physical sensations. (1987:118)

This same method characterizes the work of the skillful blacksmith and other actors (see Hill and Plath 1996; Jordan 1993). In what follows, while we will emphasize visual imagery, our discussion will incorporate kinesthetic and aural imagery wherever possible.

Blacksmithing imagery

The imagery associated with blacksmithing includes abstract mental representations of principles and mental schemata tied more closely to experience. Taken together, these constitute a major portion of the stock of knowledge. In addition, dynamic task-specific plans and constellations created and revised in the process of accomplishing particular projects are constructed as multimodal images. The two sets of images are not unrelated, for task-specific representations are informed by the repertoire of general-purpose images stored in the stock of knowledge. In turn, the generally applicable notions of form and process have been sedimented over time through repeated relevance in the experiences of specific work projects. There is an ongoing flow from general notions to specific formulations and back to revise or reinforce the stock of knowledge.

Our discussion below addresses both the stock of imaged knowledge that selectively serves as background in most forging activities and the dynamic imaging that occurs in the process of acting. We turn first to a discussion of significant sets of general-purpose images constituting the stock of knowledge.

General-purpose imagery

Fuel. As we discussed in Chapter 2, the fire is basic to the blacksmith's work. The fuel of the fire and its by-products are divided into four categories distinguishable primarily on visual grounds.

Green coal is the fuel with which a fresh fire is started. It consists of heavy, black, shiny, hard, angular individual pieces that ignite with difficulty and produce quantities of smoke and soot and relatively little heat. As volatiles are driven off by the heat, the green coal is converted to coke, which is lighter, gray, and dull, and forms large, relatively soft, solid masses through which air cannot pass. Coke is easier to light than coal, burns with little smoke, and produces intense heat. Coke is the fuel with which almost all forging is done.

As coke burns it produces ash and clinker. Ash is light, fine, dark gray material with a powdery consistency. The lighter portion of the ash is carried up the chimney of the forge, while the heavier portion collects in the bottom of the fire pot along with clinker. Clinker is formed by silica present in the coke as an impurity; in a molten state, clinker accumulates in the bottom of the fire pot, cementing together some of the ash and any other solid impurities present. It is heavy, light gray, shiny, and hard, and forms a vesicular, gelatinous mass that impedes the air flow in the forge and also operates as a heat sink, robbing the remaining coke of much of its effectiveness.

These images of coal and the stages through which it passes in burning are held constantly in mind as the blacksmith works. With them, he creates a fire of the temperature and size desired through a process of constant monitoring and matching of the perceived fire with images of what is desirable. This in itself is a demanding process that CK notes required all his attention in the early days of his apprenticeship, thereby hampering his ability to concentrate on the actual forging of a product in iron. In learning the craft, the apprentice must focus his thoughts on constructing images associated with the basic requirements of the activity, and this may interfere with the primary task at hand. As general-purpose images are learned, however, they become taken for granted or tacit and allow the smith's attention to shift from monitoring the fuel and fire to the production of an artifact.

Fire. Once the smith has acquired competence in building and maintaining a fire, the concentration required for this activity lessens and other aspects of forging can be simultaneously contemplated. But the quality of the fire is never out of mind. When a new fire is built using green coal, the smith must wait until a quantity of coke has been formed and the resulting mass is broken up into nut-sized pieces; only then can iron be heated in the fire. The decision regarding when the fire is ready for use is made on the basis of the visual appearance of the fuel.

As work proceeds, the smith must constantly monitor the state of the fire in relation to the procedure in progress. Coke is constantly burned up and new fuel must be pushed in with a poker to replace it. Sprinkling water from the slack tub helps speed the conversion from coal to coke, controls the tendency of the fire to spread, and dampens flames, which contribute to the often uncomfortably high temperatures at the forge. Decisions about when and how much water to put on the fire are made by comparing the current conditions with a visual image of the desired condition.

The temperature of the fire is gauged by color, radiant heat, and size. Decisions about adjusting the amount of air being supplied to the fire are based on these criteria and the requirements of the current procedure. A tightly packed, hot fire with the stock visible through spaces between pieces of fuel is desired for welding, while a slower fire in which the stock is covered with fuel is preferable for soaking a large piece of iron to heat it uniformly. Visual and tactile images, including the feel of the stock and the fuel, are relevant in maintaining the desired fire in these cases.

As work progresses and fuel is consumed, it is necessary occasionally to look deep into the fire to determine if there is sufficient fuel to produce the required heat and to consume the oxygen introduced by the air blast from the blower. Since the fuel tends to form a solid mass as it is converted from coal to coke, a tube of coke sometimes forms immediately above the orifice in the bottom of the fire pot. This condition is known as having a "hollow fire" and is undesirable because the temperature produced is lower than expected by other indicators and the excess oxygen causes pitting on the surface of the iron. If such a tube forms, it must be broken by pushing more coke in from the sides of the fire to fill the hollow. Presence of a hollow fire can be discerned visually and also by the characteristic sound such a fire produces. The accumulation of clinker in the bottom of the fire can be determined visually by the smith and it can be removed with the poker or shovel. Here visual and aural images are used to determine a course of action.

The smith constantly monitors the fire during forging, compares it with his image of the desired state, and makes appropriate adjustments to bring the two in line. All of this management of the blacksmith's primary tool co-occurs with and crucially enables production of works in iron. Yet this imagery is only one among many of the background sets of conceptual information that must be mastered in ironworking.

Color as a temperature indicator. One of the qualities of the iron itself that is visually salient during forging is its color. The various colors through which iron passes as it is heated have been described in Chapter 2. The smith relies on his images of those colors so that a work piece can be removed from the fire at a temperature appropriate for an intended procedure. Two pieces of iron to be welded together must have a pale yellow color and the surfaces of the material must look wet if the weld is to succeed. "Lemon yellow slick" is the way one smith has described the appearance, although attempts to verbalize these images are notoriously idiosyncratic. A bright orange heat is desired if the goal is to produce considerable changes in the shape of the stock, while a low red or black heat is used for planishing the surface of a nearly completed piece. Reference to Figure 2.6 will give the reader an idea of the range of colors each blacksmith must represent to himself for use in forging.

The critical nature of applying these visual images becomes apparent when a smith has to work under lighting conditions different from those he is used to and the perceived colors of the iron vary from the standards he has visually memorized. The difficulties many smiths encounter when doing demonstrations outdoors in the bright sun exemplify this.

Tools and procedures. Perhaps the richest inventory of images retained by the blacksmith is the set representing tools and procedures. This is a significant component of the blacksmith's stock of encyclopedic knowledge (Sperber 1974). Unlike image schemata of definitional knowledge described for linguistic categories (Lakoff 1987; Rosch et al. 1975), encyclopedic knowledge entails rich representations of details of form and process that serve as resources in production.

One knows about one's tools, where they are kept, how they are shaped, the past productions in which they have been utilized to effective or ineffective ends, from whom they were purchased or how they were made. This imagery is rich and detailed, its comprehensive quality often seemingly excessive, yet the information represented allows the smith to search his stock of knowledge with respect to the needs of a task at hand.

As we have said earlier, tools are not seen simply as members of categories but rather as clusters of features or attributes, each of which may have more than one potential use. Some tools are more complex than others, that is, they have a greater number of features than others. The simplest are the punches, drifts, and mandrels that the smith has made for himself. By contrast, hammers usually have two differently shaped faces, each of which can be used in a number of different ways, and a vise possesses multiple shapes and potential uses.

When a process is envisioned, the images of tools available are examined to see which, if any, possess the features necessary to carry out the planned procedures. As Kosslyn (1980, 1987) points out, humans have the ability to visually separate an object into its components. In addition, we are referring here to the ability to see in the tools or their components the potential for performing familiar and novel processes.

In Chapter 2 we observed that mutability is the essence of blacksmithing and that the blacksmith works within the potential for reconfiguration in his technology, strategies, and production simultaneously. An approach of this nature requires a particular kind of imagery, one that allows for the recognition of potential in objects or procedures previously used for different purposes. In an example of this, one of CK's mentors once left the anvil in the midst of a forging operation and began to wander around the shop holding the still-glowing stock in one hand and muttering, "I need a shape." He inspected and passed by the various tool racks and tables but stopped at the vise. After looking at it intently for a few seconds he reheated the work piece, returned to the vise, and forged the work over the ball on the end of the vise handle (see Figure 3.2).

In this instance the blacksmith was able to reflect on his image of the vise, which was brought to mind by his visual inspection of the shop. He deconstructed the imagery and imagined the ball at the end of the handle serving instead as a stake in his current production sequence.

A smith matches visual images constructed in seeking a resolution to a particular problem with the contents of his shop; this involves perceiving tools and equipment not as monolithic units with a single function, but rather as an assembly of components, each of which may incorporate features useful for some particular operation. A smith must always be able to transcend definitional or prototypical schemata on the basis of task-oriented criteria of relevance. This capacity suggests a rich encyclopedic knowledge of objects and procedures open to multiple applications. It is as though one's images were constructed as resources providing affordances (Norman 1988; Gibson 1977) for as yet unconceived problems.

Accompanying his imagery of tools, the blacksmith also constructs a set of images of processes representing procedural steps in production. These typically involve kinesthetic as well as visual imagery. Processual images represent successive changes of form and the actions, tools, and materials required to achieve those forms. Such multimodal imagery is reviewed in detail in construction of a plan and potentially revised in the service of production.

Images of established procedures are constructed and may be revised in practice. For example, Sanders (1994), a contemporary Midwestern blacksmith, describes an instance in which he was able to apply his imagery of making fireplace shovels to the production of chain links. The procedure he normally used for making chain links required that a rod be bent into a U, the ends scarfed or beveled on one side and then joined by forge welding to make a link. Picking up a piece of scrap rod that had been cut from a larger piece at a 45° angle, he was reminded of the process for making a loop handle for a fire shovel. In that process, the angled end allows the rod to be bent into a loop and welded to itself without the need to forge a specially shaped bevel (Sanders 1994). This recollection suggested to the smith that if he simply cut his stock for making chain links from the parent rod at a 45° angle he could avoid the beveling step and so speed up the whole process.

Here an image occurs to the smith, unbidden, but in response to the shape of a piece of scrap at hand. He recognized the relevance of details of the imaged production of an entirely different artifact, the fire shovel, to the making of chain links, and this enabled him to revise what had been his standard practice for making one kind of artifact by generalizing from the recalled imagery associated with the production of another.

The change in practice that occurred here was relatively simple,

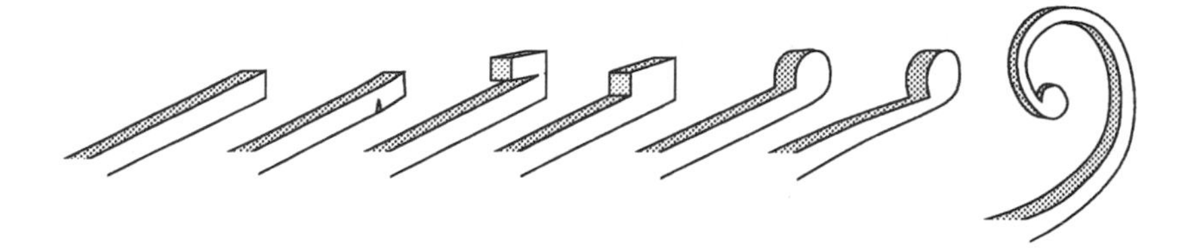

Figure 6.1. Procedure for forging a snub-end scroll, illustrating the formation of the rectangular set-up.

but the example is complex. Not only does it illustrate the importance of detail in imaged representations, but it demonstrates the potential for revision of a production sequence by reference to images of which a smith is fortuitously reminded through association with some aspect of the material conditions of work in progress.

Set-up images. Set-up images are another relatively stable although expandable component of the blacksmith's visual repertoire in the stock of knowledge. These images represent stages in production. Set-ups are goals for accomplishment in the intermediate stages of making an artifact and represent forms required to enable effective production of a final desired end. These images are typically standards learned through experience. The smith carries in his head an inventory of set-up images for particular projects and periodically draws on these images during production to evaluate work at hand and decide whether he is ready to advance to a subsequent step or whether more work at the current stage is necessary.

During CK's apprenticeship a frequent answer to the question, "How do you make an *X*?" was, "Well, the set-up for *X* is *Y*," and the shape of *Y* would be sketched on the ground or another convenient surface. In the example of the snub-end scroll (see Figure 6.1), the set-up is the partially cut end that has been bent back and welded. Forging and welding the bar in this way results in a preliminary form that allows the forging of a snub end quickly and conveniently. If a series of scrolls are required, the set-up makes it possible to forge them consistently so that the finished scrolls will read as uniform. An alternative procedure would be to upset the end of the bar and then to shoulder it and forge it round. However, upsetting is a time-consuming procedure and difficult to control. Using the set-up described is generally preferred.

In the work of a blacksmith cutler, the shape of the forged blade prior to grinding is governed by a set-up image. The blacksmith does not aim directly to produce the finished shape, although he also holds an image of the completed blade in mind; he aims rather to produce a shape that can be subsequently ground to the desired final form represented by the umbrella plan.

In the case of the production of a knife, a number of additional prior set-ups are also used. The tang is forged and the blade rough-forged to produce a rudimentary point and a back-to-edge taper. None of these has precisely the form it will have in the finished knife. They are intermediate goals that allow the smith to proceed to the final form with the maximum degree of certainty and minimum effort (Keller 1994).

A slightly different example of this kind of imagery involves using stock that has a form approximating the desired end for a project before forging begins. During a phone conversation CK asked one of his mentors about how best to make a cup to hold a candle in a candlestick. The one-word response, "Pipe," facilitated the construction of an image that conveyed all the necessary information to solve the problem. The cups were subsequently made by forging them on the end of a piece of pipe, sawing them off, and torch-welding them in place. A single word in this case allowed CK to construct an image of a form that he used as a set-up for achieving a goal and then to reason visually about the procedures for attaining this goal using the set-up proposed.

Aesthetic standards of form. In addition to the imagery specific to forging, artist-blacksmiths also employ imagery from other contexts. Perhaps most notable among what we might refer to as imported imagery are criteria for the evaluation of forms. Aesthetic criteria

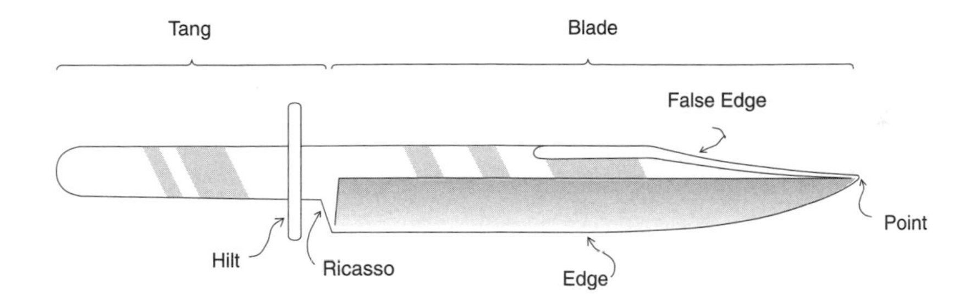

Figure 6.2. Outline of knife blade showing primary components.

such as straightness, evenness, and fairness are schematically represented and applied in a wide variety of contexts for the production of ironwork.

The use of the concept of straightness can be illustrated in the work of cutlers. As a blade such as that depicted in Figure 6.2 is tapered from ricasso to tip in the forging process, hammer blows must be alternated between the two faces of the blade to keep it straight. This process is monitored by reference to an image of straightness.

In the process of forging any linear shape the work may tend to curl up away from the anvil if the hammering is confined to only one face. If the face of the hammer contacting the work is not parallel to the anvil's face but is canted to one side or the other, the work will tend to curve in the opposite direction as one edge of the material is compressed more than the other. Consequently, the end of a forging episode is often marked by the smith lifting the work, sighting down its length, correcting any curves or bends, and then returning it to the fire.

In addition, it is not uncommon to see a smith, who is inspecting the finished work of another, unobtrusively sight down the long axis of the piece to see if it is straight. Straightness is a valued feature in hand-forged work, but it refers to "reading as straight," as discussed in Chapter 2, rather than an absolute property checked against a mechanically produced gauge. The piece in question, whether finished or in process, is compared to a visual image of straightness and evaluated accordingly.

Other similar aesthetic characteristics include evenness or consistency, as in a twist; fairness, as in a curve or scroll; smoothness; symmetry; and uniformity, as in a taper. Such images are skeletal forms, not unlike the imagery that Johnson (1987) and Lakoff (1987) argue provide a source for visual understanding in many domains.

Sometimes images of aesthetic standards are somewhat more specific. While an apprentice, CK wanted to make a large spoon. The familiar frontal shape of a spoon was produced with some facility. The result was a straight piece with appropriate cross-section changes that needed to be given the curve characteristic of a spoon when viewed from the side. To his chagrin CK could not remember what a spoon looked like from that perspective, and could not find one in the shop for reference. The advice from one of his mentors was, "All spoons are S's." A series of S's were sketched ranging from horizontal to vertical and covering the variation from shallow table spoon to deeply curved ladle.

Aesthetic criteria and set-up images can be distinguished from the richer encyclopedic imagery for fuel, fire, other tools, and procedures. Aesthetic criteria and set-ups are primarily applied as standards for production or constraints on outcome. Encyclopedic imagery plays an additional role as a resource for planning, rearrangement, synthesis and novelty.

Task-specific imagery

We turn now to a discussion of the construction and dynamic application of imagery in production.

The umbrella plan. Conceptualizing a goal for production is the initial stage of making something in iron. As we discussed in Chapter 5, the blacksmith typically formulates a mental image of a desired product, which may then be graphically depicted as in Figure 6.3. Such imagery we refer to as the umbrella plan. The plan implies means for accomplishment in addition to outlining a desired form. The importance of the imaged forms and procedural entailments of

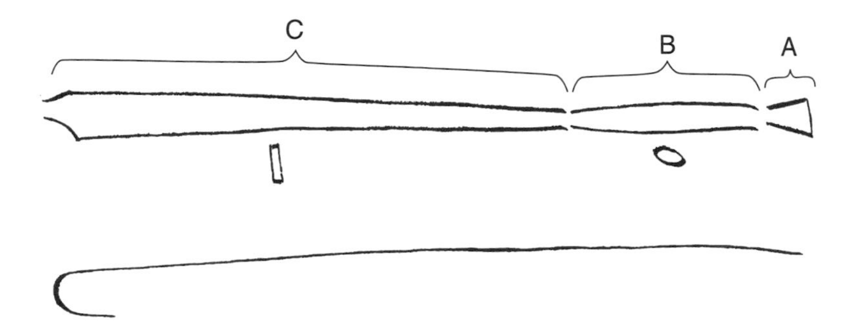

Figure 6.3. Design sketch of 19th-century-style skimmer handle, drawn by CK prior to initiating production.

an umbrella plan for successful production was stressed by a blacksmith cutler in an interview with JK: "I have to get a concrete visualization of the knife going, because I'm going to be out there shaping it with the hammer. I'm dealing with dimensions."

Imaging the form allows the smith to envision the steps for production and thus comprehend processes for realizing the form. The images constructed for production may range from skeletal compositions to more detailed representations, but they rarely if ever provide exact specifications. Instead they provide an orienting plan and serve as a mutable guide against which progress in material production is evaluated.

Certain features of an umbrella plan are more specific than others and less subject to change in the process of production. For instance, the size of the finished object must be known fairly precisely for the proper size of stock to be chosen. Other dimensions of the umbrella plan also lead to initial decisions not subject to later changes. For example, the general strategy of forging operations to be used in production will influence the shape of the stock to be selected. Once selected, the raw material for a project is not replaced unless the entire process is initiated anew, although additional segments of stock can, if necessary, be welded to an original segment of iron to increase the available mass or length. The dimensions of the raw material are re-formed in production, but characteristics of the original stock may have enduring implications for both procedure and formal results.

Other features of the umbrella imagery, however, are more open to construction. Salient design features that define the style or category of the object must be included initially in an umbrella plan, but their specifics may be vague. An early 19th-century Pennsylvania German skimmer would have a rat-tail finial, but the direction of its curve can be decided upon as the work progresses.

The significance of the umbrella plan for acceptable behavior is nowhere more clearly illustrated than in making the decision that a project is completed, or that work on a given piece must be abandoned, or, more rarely, that the image of the goal must be significantly modified to conform to the object produced. Such decisions are based on a comparison of the product with the image of the desired goal held at the outset of the project and modified by whatever exigencies have arisen in production. This can be a process of gradually approximating, in iron, a desired goal as imaged. More often the process entails bringing both image and ironwork closer together in the emergent and dialectic process of forging. The matching of image and product may be complicated, for how close the match must be is often influenced by other factors. The correspondence between an image of the goal and the object in production can be conceptually assessed as good enough if, for example, investing more time in the project is not feasible, or if heating one part to forging temperature one more time might endanger the entire piece.

As argued in Vera and Simon (1993:17), "Most plans are not specifications of fixed sequences of actions, but are strategies that determine each successive action as a function of current information about the situation." While we would disagree with the determinism indicated here and prefer to see strategies and current information as resources for successive action, we applaud the notion of plan as strategic and situationally adaptable. An umbrella plan must remain open to the potential for increasing specificity and reformulation as a blacksmith works. The process is dynamic and dialectic. Images govern production, the material results of which, as perceived, allow revisions and refinements to the governing images. This is the essence of emergence.

Constellations. As we have argued in Chapter 4, constellations are the constructs that make productive acts possible. Guided by the umbrella plan, the blacksmith integrates mental expectations and material elements in an approach to the accomplishment of a particular procedural step in a sequence of one or more steps anticipated for making something. The constellation has an immediate goal for transforming the work, implies a particular procedure to be enacted by the smith, and requires materials in hand and tools in reach. The constellation in its mental half integrates visual, sensorimotor, and aural imagery. Figures 6.4, 6.5, and 6.6 depict some of the visual images that might be part of a constellation for making the snub end of a snub-end scroll.

Governing a constellation for production of a snub end for a scroll is a conception of the completed element. The imagery of a constellation, then, is a potentially mutable set of representations of a desired end, procedures, and tools. The sequence of forms represents the set-ups and implies the procedural actions for the production of the snub end on a length of iron stock.

The raw material – stock of certain dimensions – selected for the work and the hammer chosen for this project figure in the imagery depicted in Figure 6.4. A wire brush and flux necessary for welding are added in Figure 6.5. The anvil is differently highlighted in the imagery of each constellation as respective dimensions of this tool are relevant to a particular step. The fire, controlled for the proper temperature, quality, and state of the fuel, is also uniquely represented in each of these constellations.

As this operation might proceed, the various pieces of equipment or material are collected from the shop. For instance, a visual image of the shape needed in a tool will orient the visual perceptual survey of the inventory of tools and lead to selection or manufacture of an appropriate implement. Selected tools and materials will be used once or repeatedly as necessary, guided by the images of the constellation.

Setting to work with the constellation in Figure 6.4 guiding him,

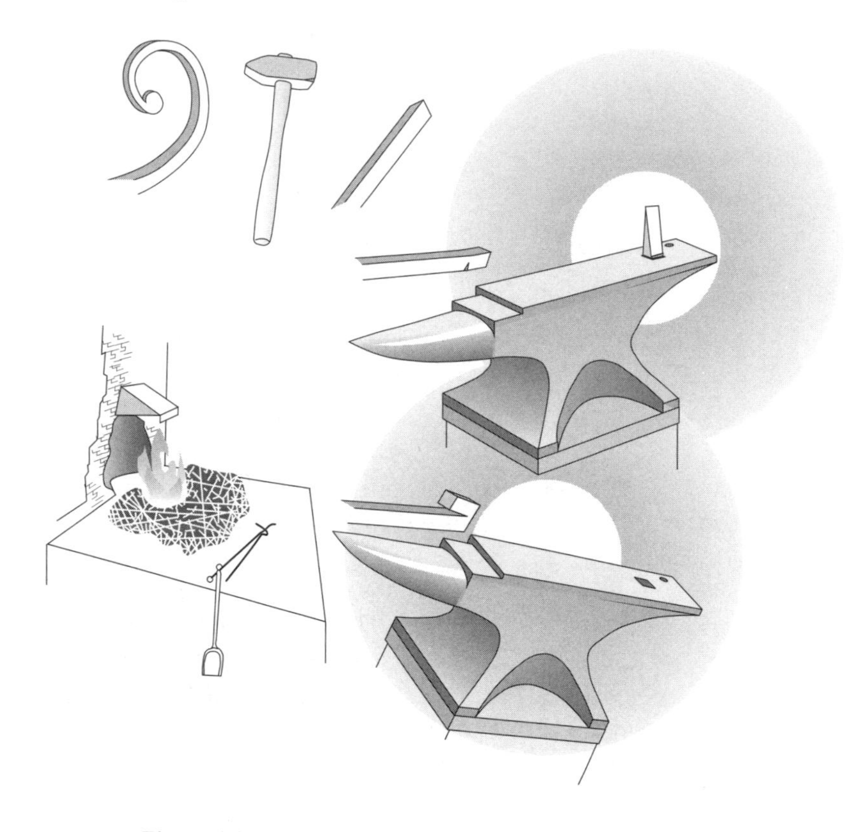

Figure 6.4. Initial images arranged in a constellation for preparing the set-up for production of a snub-end scroll. The first steps in production of a rectangular block are illustrated, with the ultimate goal in mind in the upper left and the shape prior to welding shown between the two anvils.

the smith heats the stock and cuts part way through near the end, and bends that end back against the bar. Visual and kinesthetic images are applied in the process of cutting and bending the stock. For example, the end will fall off when bent if the cut is too deep, but an excessively shallow cut will not permit the end to lie flat against the bar after bending. It is both the visual information and the feel of the depth of the cut that the blacksmith uses to achieve the desired result.

Figure 6.5. Images in a constellation for welding in the making of a snub-end scroll. The ultimate goal appears at bottom. Tools and the rectangular form after forge welding are depicted above.

After the bending is complete, the bar is reheated and wire-brushed to remove accumulated scale. Color imagery is used here to insure that the bar is hot enough for the scale to be removed by brushing. Then the bar is fluxed and returned to the fire. Color imagery is significant here as well. If the bar does not retain a nearly orange heat after brushing, the flux will fall off instead of sticking and melting when applied. Too much cooling, reflected by a color

Figure 6.6. Mental representation of the final steps in the making of a snub-end scroll. The ultimate goal is represented on the left and the final outcome on the right, with intermediate forms and tools placed between them.

change during wire-brushing, would occasion reheating before fluxing.

Next, the work is brought to welding heat and the end welded in place against the bar as represented in Figure 6.5. This process requires visual, kinesthetic, and aural imagery. The fire must be checked and adjusted if necessary and the air blast increased as required to bring the work to bright yellow, a welding heat. The fluxed bar and bent-back end must be rotated in the fire so that both portions come to welding heat at about the same time. As the correct temperature is approached, the smith glances at the anvil to see that it is clear of scale from earlier steps and that the hammer is within easy reach on the anvil face.

Subsequently, to weld the pieces together the smith must understand what the necessary light, rapid blows feel like. Blows too light, too heavy, or too slow will fail to make the weld. Blows that are only a little too heavy will make the weld but distort the shape of the piece.

After wire-brushing again to remove scale and flux residue and after reheating, the square lump on the end of the bar is forged into a rounded shape, the curve of which will complement the curve into which the bar will be scrolled (see Figure 6.6). At this point, with recourse again to the pertinent visual and aural images, the fire and blower are adjusted so that a forging rather than welding heat will be generated. Visual and kinesthetic images are used to position the square lump appropriately on the anvil as the corners are pounded down, the resulting side projections are flattened, and the square is transformed into a circle. This cycle of forging corners and flattening sides is repeated several times until the work conforms to the (mutable) visualization of the desired shape.

As this example illustrates, enacting a constellation and completing a step in production are dependent on the ability to match images with a current state of a product. The imagery drawn upon for evaluation may be both general-purpose and task-specific. General-purpose images provide background reference to common forms and processes. Task-specific imagery provides details of shape and procedure employed and enriched in a particular production.

Failures to achieve an acceptable match of image and work can result from a number of different circumstances, including lack of skill and distraction. More complicated, however, are those failures that result from attempts to execute a constellation inappropriate for production of a desired form. One smith has spoken of the floor of his shop being "littered with my failed attempts to do something." Another remarks, "I worked for two days on the thing and it still wasn't right. I hadn't made any mistakes, I was just going about it wrong." In these cases the procedures envisioned as resulting in the goal imaged proved to be incorrect ones and developing a new constellation was necessary before the project could be completed.

General remarks

The preceding discussion demonstrates that nonverbal imagery is a significant element of the stock of knowledge and as a mutable standard for production constitutes a significant component of the conceptual thought involved in emergent practice. Integral to the knowledge and cognitive capacities involved in the work of contemporary artist-blacksmiths are an inventory of images, the ability to match images with direct perceptions (see also D'Andrade 1991), and the ability to generate new images from the components of prior visual representations and present perceptions. These characteristics and capabilities richly substantiate the structural and functional analyses summarized at the beginning of this chapter and go beyond them in providing new detail regarding the place of imagery in the conceptual thought of creative and practical activities.

Imagery in the stock of knowledge

Imaged schemata constitute a primary component of the stock of knowledge. The blacksmith acquires and constructs numerous images and this inventory is infinitely expandable. The motivation for acquiring and maintaining the many necessary images for productive activity comes from the goal orientation and practices of an individual. Images are constructed and preserved – sedimented in Schutz's terminology – for their potential value in anticipated activities and are recalled, sometimes unbidden, as new and related activities take place. This is knowing for doing. Sedimenting images through practice continually builds the artist-blacksmith's stock of knowledge, enriching the possibilities for actions, which, in turn, enrich the stock of knowledge.

Imagery in production

Throughout this book we have described the productive process as an emergent one. Using the stock of knowledge as a resource, the blacksmith develops a plan and hypothesizes means to his ends. The umbrella plan itself may be an emergent design. One blacksmith discussing his practices refers to flipping through large numbers of images in search of those relevant for constructing a novel umbrella plan. Once productive activity is initiated, what remains to emerge is the final material result, the ultimate procedural sequences for attaining that result, and the detailed image of the outcome. This emergence involves a continual interplay of actions, objects, and images, each potentially affecting and affected by the other.

The process of production entails constant monitoring of results so that progress is recognized and problems in the appearance of a piece of work are diagnosed. This requires holding a multitude of images in mind and comparing perceptions of the material with relevant images so that appropriate action can be taken. As we mentioned in Chapter 2, CK's mentor repeatedly criticized failure to monitor the temperature of the stock being forged, as indicated by its color and malleability and as produced by the state of the fuel. Eventually CK learned to hold all of the sets of images simultaneously and to shift between them fast enough so that a task could be accomplished appropriately.

Where images are vague, progress toward a goal may entail producing an inappropriate material result, which in its very production contributes to clarifying the desired characteristics of the final product. Images must be enhanced as a project proceeds in order to anticipate details of procedure or form not envisioned in advance. What ultimately allows activity to cease is a judgment on the part of a smith that an image of the desired goal and the actual forged product are adequately matched.

While it may be difficult to test these claims definitively, clues to their accuracy may be found in the smiths' reflections on their work. For example, asking someone how they know when their work is done elicits answers such as "Because it looked right" or "Because it worked." Replies like these express satisfaction in the correspondences between material results and images. Simply completing a sequence of operations is not sufficient for a smith to conclude that a product is finished. Judgments of an acceptable match between images and material are essential.

Skill and imagery

We wish to stress the conceptual role of imagery as the foundation for skillful action. Skill is as much a manifestation of acuity in visual representation, facility in visual manipulations, and diversity in the repertoire of constructed nonverbal images as it is a reflection of hand–eye coordination. As we discussed in Chapter 2, skilled performance in any activity requires a sensitivity to the effect of any given operation on a piece of work. This is primarily an imaged projection of a future state based on the envisioned enactment of procedures for its attainment. In building a repertoire of images, the apprentice increases his sensitivity to and subtlety in the application of the images.

Gladys Reichard (1974 [1936]) mentions that experienced Navajo weavers quickly realized when excessive weft tension was pulling a partially completed textile away from its intended rectangular plan into an hourglass shape. On the other hand, Reichard herself as a novice weaver often had to rip out quantities of weft and reweave to correct for constriction unnoticed in its early stages. The ability to compare the results of an enacted procedure with an image of what an acceptable result should be is the touchstone of effective production (Jordan 1993; Hutchins 1983; Hill and Plath 1996).

Imagery and reason

Umbrella plans and constellations, which are ways of organizing prior knowledge for the production of ironwork objects, are the results of a largely visual reasoning process. Revisions to these plans or hypotheses and decisions regarding acceptability of an outcome are also achieved largely through nonverbal reasoning. This thought takes advantage of the structural capacities of the visual system, as noted by Kosslyn (1981, 1987) and Marr (1982). Images are compared, components rearranged, orientations altered, and projections constructed in the mind. As a consequence, an adequate account of the knowledge underlying appropriate behavior entails recognition of the set of images employed in the practices of interest. It is often the images themselves that embody the crucial knowledge enabling skillful performance.

Field work

One profound implication of these conclusions for anthropology and perhaps the cognitive science disciplines more generally is that one needs to practice an activity in order to represent knowledge appropriate to it. Images are not easily externalized for public consumption. Nor is it a straightforward matter to translate imaged ideas into verbal renditions. Observers and learners of an activity refer to the difficulty craftspeople have in articulating a full and rich description of their work. In these cases it is often the nonverbal imagery that remains ineffable. We would argue that nonverbal imagery can best be acquired and integrated with other relevant knowledge through engagement in a set of practices at issue. This suggests that the scholar interested in knowledge should take seriously the participant half of participant observation (see also Hill and Plath 1996; Jordan 1993).

In summary, a skilled artist-blacksmith forges a product in iron by visualizing a goal, seeing the procedures for its creation in an imaged form, and acting on that plan with a mind open to alterations of the image and innovations in form. This process requires handling multiple images in rapid succession: the fire, the plasticity of the iron, the feel of the hammer, the expectations for shape. The process also requires maintaining the goal-state images while perceiving intermediate changes in the empirical state of events associated with production. Handling these multiple images is difficult for a beginning smith but becomes easier as experience is gained and, perhaps, images become ingrained. Still, there seems to be a limit to the facility with which one can handle the complex imagery required for production. While it is possible and often desirable to have more than one length of iron for a given project in progress at one time, it is very difficult to have more than one project under way at the same time. While the limit for each artist-blacksmith is probably unique, the juggling of a large number of unrelated mental images at some point leads to distraction and confusion rather than effective production – hence the traditional warning against having too many irons in the fire.

7 Beyond blacksmithing

Accounting for productive social activity requires attention to cognition and practice as cultural and individual phenomena. An anthropology of knowledge provides such a focus. This approach assumes that knowledge governs practices, which in turn reproduce or revise prior knowledge. It further assumes that shared ideas, products, and events, which constitute culture, are constructed in this dynamic process. The process entails anticipation and accomplishment. It is the foundation for and expression of the principled creativity and customary routines that characterize human behavior.

Knowledge is not a static body of information. As John Seeley Brown described it (public lecture, 1995), knowledge is knowing at rest, open and diverse systems of conceptual entities and their relations. Knowledge is organized as relevant to empirical domains of experience and to accomplishment within these domains. When fully elaborated, such a system includes governing principles and information about procedural possibilities, instruments, artifacts, and social, aesthetic, and functional relationships. This stock of information is largely represented in schematic form as generalizations from experience with visual, kinesthetic, and propositional dimensions.

A knowledge base serves as a resource for the construction of plans, including the umbrella plan and hypothesized means for attaining envisioned ends. Constellations of implements, materials, and ideas enable the testing of hypotheses in practice. The material results of action may lead to confirmation of an original plan and associated hypotheses or to revisions in these prior conceptions. This constitutes the emergent dimension of production.

Once an event is completed, an image or description of the outcome and information drawn from the production sequence may be incorporated into the stock of knowledge available for future consideration. In this way a knowledge system continually adds new information to old. Yet over time, domain-specific knowledge is also increasingly coherent as principles are formulated and densely integrated with practices. This principled coherence gradually creates conservatism in the foundational ideas and boundaries of a system of knowledge. A productive tension results from this tendency toward conservatism, which contrasts with the open quality of the stock of knowledge.

This framework has allowed us to posit principles shared among artist-blacksmiths defining their domain of activity, to identify the stock of knowledge centrally relevant to and derived from their practices, and to account for the productive process and its material outcomes. While the previous pages are an analysis of artist-blacksmithing, we wish to demonstrate now that this is *not* a book about blacksmithing. In this last chapter we offer evidence from the work of other scholars that suggests a wide application for an anthropology of knowledge to diverse domains of experience.

One of the closest parallels to our own work appears in research by Carlson (1991) and by Gorman and Carlson (1990, 1992) on understanding invention as a cognitive process. Using "notebooks, sketches and artifacts produced . . . by Bell and Edison," these scholars aim to reconstruct the mental processes involved in the development of the telephone. Adapting concepts from cognitive psychology, they suggest that invention is an emergent combination of ideas and objects. The former they refer to as *mental models,* the latter as *mechanical representations.* Strategies or heuristics for problem solving enable inventors to bring mental and material components successively into a closer and closer approximation of one another.

While their vocabulary differs from ours, the framework is remarkably similar. Inventors' mental models are typically imaged representations and often take the form of an umbrella plan projecting a goal for production with procedures or heuristics for attaining that goal. Gorman and Carlson state that mental models are frequently "incomplete and unstable and an inventor often mentally considers competing arrangements" (1990:136). These qualities also characterize the umbrella plan. The skeletal nature and open quality of such orienting conceptions are as crucial to accounting for the activity of invention as for that of blacksmithing.

Governing principles also appear in Gorman and Carlson's analysis of invention. They note that "an inventor possesses a mental model that incorporates *his or her assumptions* about how a device might eventually work (Gorman and Carlson 1990:136; emphasis added). These assumptions can be more or less abstract but they transcend any particular model. In the case of Bell the enduring assumptions were two:

> First, Bell saw the telephone as a device that converted complex sound waves into a fluctuating current. Second, Bell also knew that he needed to express this general principle in physical terms, and consequently, he sought a physical phenomenon that converted sound into electricity. (Gorman and Carlson 1990:145)

Electromagnetic induction was the mechanical principle he returned to again and again in his experiments. For Edison, by contrast, "variable resistance served as the central idea . . . of his experiments with the telephone."

These general principles, and perhaps others difficult to reconstruct from historical events, characterized the experiments designed by each inventor over many years. Their respective approaches to developing a telephone diverged as a result of the principles each held as a foundation for his work.

The dynamic process we have outlined for artist-blacksmithing also seems characteristic of the inventor's production. Following the original conception of a goal, "the essence of invention," Gorman and Carlson argue, "is the dynamic interplay of mental models with mechanical representations" (1990:159). The inventor conceives of a possible device and tests his ideas in material models. Like the blacksmith, the inventor possesses a "mental set of stock solutions," a stock of knowledge on which he may selectively draw or which may be selectively reconfigured as problems arise in the construction of mechanical representations (Gorman and Carlson 1990:141).

When the solution to a task at hand is lacking in the stock of knowledge, however, the inventor relies on the performance of a mechanical device to provide critical information for continuing experimentation. For example, in working on a telephone, "experiments with mechanical representations led Edison to revise his mental model" (Gorman and Carlson 1990:154). He assumed at one point that variable resistance could only be achieved "by moving a needle through a high resistance material, but . . . experiments showed him that variable resistance could be secured by directly vibrating the resistance material" (Gorman and Carlson 1990:154).

Gorman and Carlson point out that in attempting to transmit acoustical vibrations, Edison first relied on an intermediate substance (initially a needle but later other substances were tested) to transfer a signal from an iron diaphragm to a high-resistance carbon button. Failure of intermediate mediums to conduct the vibrations adequately, however, ultimately led Edison to hypothesize that direct transfer of the signal might be effective. This insight was derived from the unintended material results of Edison's experiments. Edison subsequently constructed a device based on this insight. The success of that device reinforced the new principle of transference and the new mechanism became the standard configuration for the carbon telephone transmitter.

Although Gorman and Carlson do not directly refer to a unit of mental and material elements enabling a step in production, a *constellation* in our framework, this notion seems relevant to an account of invention. Carlson (1991:155) points out that Edison devoted considerable effort to the organization of a new laboratory in the late 1880s. We have noted that the blacksmith's shop is constructed and organized to reflect the work of forging and to make tools and metal visually accessible. Similarly, Edison's laboratory, Carlson argues, is "an artefact of the invention process" (1991:150). "Throughout its history, its arrangement was in constant flux, with Edison skillfully adapting it to the inventions underway" (Carlson 1991:155).

The notion implicit here is one of gathering appropriate tools and raw materials to be accessible for experimentation. " 'Everything from an elephant's hide to the eyeballs of a United States senator' " (Carlson 1991:155, quoting newspaper reports of the day) was represented in the laboratory's store. Edison set up distinctive shops for electrical and chemical work, for carpentry, for mining, and for metallurgy and blacksmith work (Carlson 1991:154). He hired experts in each area to conduct his experiments. It seems likely that for each project particular supplies were selected as relevant to the anticipated experimentation and that with each step in production the relevant resources would be clustered at an appropriate work space in a specific shop site.

Carlson (1991) interprets Edison's working style as that of a craftsman, a comparison that suggests the ties to artist-blacksmithing noted above. There are, of course, differences between invention and artist-blacksmithing as well. The goal of production must often be fuzzier for the inventor than for the smith and the procedural steps to achievement less apparent. The time to accomplishment of the ultimate goal must typically be considerably longer for the inventor. In addition, the working materials and physical activities of contemporary artist-blacksmiths overlap minimally with those crucial to Bell and Edison. Yet a common account of the process of making something appears to hold for both smiths and inventors.

We believe that this account may characterize many productive human endeavors. We have encountered few examples in the literature that develop a framework as close to the one proposed in our book as that developed by Gorman and Carlson. Perhaps this is not surprising, as authors present us with findings derived from their orienting questions, which have not replicated our own. Yet even given the diversity of the literature on human productivity, we do find many examples where one aspect or another of our framework is apparent. This suggests that perhaps fuller investigation from the position developed here would yield comparable accounts of the processes of making things. We draw on the reports of scholarly research for illustrations.

In Chapter 2 we described principles that contribute to a definition of artist-blacksmithing as practiced in contemporary America. These principles serve to govern the activities of the craftsmen we have worked with. A similar relationship in which underlying principles govern practice is evident in many other activity systems.

One example comes from the research of Heather Lechtman on

pre-Spanish Andean metalworking and textile production. Lechtman indicates that metalworking was governed by a principle specifying that that which appears superficially to be true of an object must also be inside it, or part of its essence (1984:30).

Lechtman points out that the colors of silver and gold were highly valued and appear "to have strongly influenced the visual manifestation of status and power." This led Andean metallurgists to attempt the production of "metallic gold and metallic silver surfaces on metal objects that were made of neither metal" (1984:15). In the Dynastic Egyptian and later the European tradition of metalworking this task was performed by covering an object of another material with a very fine film of gold leaf, originally applied mechanically. Later, surfaces of gold and silver were applied to objects made of other metals using chemical means.

These would not have been acceptable solutions in the Andean case, however, since they violate the principled "basis of Andean enrichment systems [which] is the incorporation of the essential ingredient – the gold or the silver – into the very body of the object" (1984:30). To adhere to this principle Andean metalworkers made objects of alloys containing small quantities of silver or gold and then, using a technique called depletion gilding, concentrated the desired metal at the surface of the object.

The same underlying principle influenced Andean textile production. Textile decoration was not applied to the surface of the fabric as in a European tapestry. Instead, the decoration was produced by manipulating the elements that actually constituted the fabric. "Andean weaving seems to have responded to notions that saw the achievement of [a] visual, surface message as emerging from underlying, invisible structural relations" (1984:33).

Another example of the influence of underlying principles on activities can be found in Hardin's description of ceramic production in San José, Michoacán, and Zuni, New Mexico (1977, 1984). She suggests that differences in decorative styles between Tarascan and Zuni potters may be attributed in part to differences in the underlying principles governing their work. This is demonstrated in her comparison of vessel decoration. In San José the production and decoration of pottery for nearby markets is based on several conceptual premises that at once constrain the decorative activity and yet provide considerable latitude for alternative realizations that fit within the limits imposed by the defining principles:

> The work of each painter is distinct. The outside observer learns quite readily to recognize the work of each artisan . . .
>
> Many of the distinctive characteristics of a painter's personal style stem from the fact that the design structure, which is largely shared by all the painters in the village, offers a number of alternative choices at almost every step of the painting process. (Hardin 1977:112, 114)

The configuration of the decoration on each vessel is predicated on a principle that establishes three zones or bands to be filled with basic design elements or motifs. These can be varied in combinations so that no two vessels are identical. Consequently, Hardin argues,

> potters treat decorative styles as systems of knowledge. The skillful Tarascan artist is said to "have a good memory for designs." These potters not only appreciate a complex design when they see one; they take it apart, reassemble it and modify it to suit a preconceived structure. (1984:583)

This preconceived structure entails the governing premise noted above and the basic classificatory schemes of elements and motifs.

In addition, the production of Tarascan potters is guided by aesthetic principles and standards. Decorations should be "well painted." This refers in part to standards of consistency and symmetry. Specifically, the three zones of a vessel surface should be more or less covered with painted designs and the lines used to paint the designs should be thin or fine lines.

These criteria are analogous to the transformative principle and aesthetic standards noted for artist-blacksmiths. In our discussions of artist-blacksmithing we have suggested that governing principles may be compromised as a smith practices. For example, the desire to complete a project in timely fashion may lead to a compromise in the freehand quality of the work. Similarly, Hardin notes that the Tarascan aesthetics as well may be compromised if a conflict with economic principles is perceived:

> Economically motivated decisions, like aesthetic preferences, have a pervasive effect on the whole of the vessel's decorated surface. In many cases, a painter's modification of his graphic style represents a compromise between aesthetic standards and economic considerations. (1977:117)

In summary, San José potters think of vessel decoration independently of the shape or function of a ceramic pot. They recognize zones to be filled with design motifs and smaller units of individual design elements. As a result, when asked, they can paint examples of basic design elements in a taxonomic, tabular form (Hardin 1984). These are elements to be manipulated in pot decoration. The ceramic artist is free to conceive of a decorative style and specific ornamental designs that can be applied to variously shaped vessels intended for diverse social functions.

Zuni potters, by contrast, do not think in terms of isolated design units but rather in terms of culturally specific types of whole vessels. "Zuni potters asked to paint designs out of context produced a relatively unorganized corpus in comparison to the Tarascan one" (1984:584, taken from Bunzel 1929). The reason for this is that "in the Zuni case, an indigenous system of classification that focuses on the entire vessel is important" (1984: 599). The formal, decorative, and functional characteristics of a vessel are integrated in a classification scheme that distinguishes ideal types of completed pots. For the Zuni, then, a vessel is not conceived of as a canvas on which decoration is placed. Instead, Zuni potters produce vessels that from the outset adhere to classificatory principles intimately linking vessel form, social function, and decoration in a single unit.

To shift to a very different domain, the problems of human interpretations of messages from a photocopying machine were studied by Lucy Suchman (1987). She applies a framework for the understanding of human interaction that overlaps in interesting ways with our perspective on productivity. Suchman explores the difficulties that arise in human interpretations of machine instructions, such as those available through an expert help system attached to a large and relatively complex photocopier (1987:98). She starts from the premise that "the organization of situated action is an emergent property of moment-by-moment interactions between actors, and between actors and the environments of their action" (1987:179). While her object of study is complicated by the dialogic process of human interaction, Suchman nonetheless sees emergence as characteristic of naturally occurring human events.

In addition, in accounting for failures of human–machine communication, Suchman points to the differential construction of plans in naturalistic human behavior and in engineered programs for self-help systems. Naturally developed plans, she argues, are inherently vague, a quality that ideally suits them to the expectation that the details of a goal and procedures for its attainment "must be contingent on the circumstantial and interactional particulars of actual situations" (1987:186). This is precisely the process we elaborate with the successive detailing of an umbrella plan as production of an artifact proceeds.

> To the cognitive-scientist, [however,] this representational vagueness is a fault to be remedied . . . The task of the designer who would model situated action, therefore, is to improve upon, or render more precise and axiomatic, the plan. (1987:185)

Ironically, it is in the very process of attempting to "anticipate and constrain" (1987:189) the actions of human users of machines through increasingly closed plans that a designer limits the potential for the resourceful use of information, which is the hallmark of human activity. Here it is evident that misunderstanding of the character of human activity leads designers to develop machine interface programs that unintentionally require highly specialized interpretive strategies. The challenge to design machine communications systems that take advantage of the properties of productive human activity clearly remains. As Suchman points out, "the attempt to build interactive artifacts, taken seriously, could contribute much to an account of situated human action and shared understandings" (1987:189).

Yet another comparison to our own work comes from the research of Walter Vincenti, whose goal is to "give scholars concerned with technological affairs a deeper understanding of how engineers think and how their thinking relates to their doing" (1990:12). In this effort Vincenti discusses the emergent qualities of aircraft design collaboratively developed by co-workers.

Our own description of the domain of artist-blacksmithing has focused on individuals and the creation of cultural artifacts, beliefs, and procedures that result from productive social activities. However, from Vincenti we find that the actions of a team who share a single task at hand may have many of the same qualities that we have described in the preceding pages.

These qualities are apparent in Vincenti's description of design and production in the aircraft industry (Vincenti 1990). As with an umbrella plan, a "design . . . denotes both the content of a set of plans . . . and the process by which those plans are produced." This "tentative layout," the product of one or more designers, is checked mathematically or experimentally by production engineers; the design is returned for modification when problems are discovered. This "usually requires several iterations before finally dimensioned plans can be released for production" (1990:7).

Here we find again, as with invention and artist-blacksmithing, the critical dialectic between the material results of practice and the conceptual plan for production. Goal-oriented action, whether on the part of an individual craftsperson, a solitary inventor, or a segment of an industrial concern, is characterized by a mutually constitutive dynamic relationship in which knowledge and practice are each continually derived from and altered by the other.

Vincenti also comments that "engineering knowledge grows typically through a complex interplay of experiment and theory" (1990:12) and that "day-to-day design practice not only uses engineering knowledge, it also contributes to it" (1990:232). This is the open relationship between a stock of knowledge and production that we have noted as characteristic of artist-blacksmithing as well.

Methods

The examples sketched above, coupled with the research developed herein on artist-blacksmithing, is suggestive of the potential of an anthropology of knowledge to account for a diverse range of human activities. Further research should clarify the limits on the scope of the framework. Where dialogue is a crucial component of activity, an additional account of the dialogic process will be essential. Where failure to achieve a desired outcome has severe, even life-threatening, consequences – as is often the case, for example, in navigation (Gladwin 1970; Hutchins 1983, 1993a) – the quality of emergence may vary. Yet even in these activities the process of goal attainment itself may demonstrate emergent properties while the final outcome is more constrained than in other endeavors. Refining the framework we have outlined here will lead to increasingly adequate accounts of human activity, accounts that recognize the integral and mutually constitutive relations between knowledge and practice in cultural systems and between social communities and their individual members.

Research directed toward these ends may require a convergence of methods. Unlike anthropologists who have shied away from extensive immersion in activities and social life, in part to avoid the risks entailed in "going native," we suggest that it is perhaps essential to acquire competence in the activities that constitute the subject of research if a scholar wishes to test the premises developed herein. It is through practice that the goals of activity become clear, and it is only through learning how to accomplish the goals one's subjects set for themselves that knowledge relevant to their tasks becomes apparent and the complexity of the relations among knowledge and practice in a domain can be recorded and explored.

Practicing the target subject of one's study is not a simple task, nor is it sufficient in itself as a methodology, but it may be necessary for further developing an anthropology of knowledge. In spite of the risks inherent in practice as a research tool (Lemonnier 1992:27) without the data a practicing competence provides, accounts of activity will be forever relegated to descriptions of systems from perspectives that lack the understanding of the agents who continually construct and reconstruct those very systems.

A procedure carried out by a skilled and experienced actor often appears smooth and effortless. Generalizations based on repeated scrutiny of an ongoing activity by a nonpracticing analyst emphasize the inevitability and rigidity, or sometimes the incomprehensibility, of the actions in question. To appreciate the reflected alternatives never enacted, the pivotal nature of certain points in a sequence, transitions in form or operations, the hint of impending problems, the repertoire of compensatory or corrective procedures and split-second choices among them, the investigator must draw on knowledge rooted in experiential as well as observational insights.

Acquiring competence in productive activity as a research tool must be constantly monitored by the researcher him- or herself, however, to allow continued access to principles and schemata that otherwise become taken for granted. The process must be complemented by traditional research methods including daily journal records, critical observation, interviews, and hypothesis testing designed to assess the adequacy of dimensions of the stock of knowledge posited by the researcher. A basic tenet of cognitive anthropology reminds us that participation in practice will generate contextually appropriate questions to pursue using the techniques of interview and observation (see also E. Cooper 1989). Exploring problematic aspects of activities should be particularly useful in this endeavor, for it is in problem solving that otherwise taken-for-granted dimensions of production may surface in articulate discourse.

The place of an anthropology of knowledge in current debates

The disciplines of cognitive science and anthropology today find themselves in the midst of debates that pit one intellectual approach against another and in extreme versions deny the promise of a higher-order intellectual synthesis to be derived from a complementarity of theoretical positions and comparative investigation of empirical settings. The anthropology of knowledge outlined herein grows out of an integration of diverse intellectual positions; this integrated approach has led us to appreciate the potential for new positions and new understandings to arise from complementary approaches and comparative investigation.

One form this debate takes in the interdisciplinary field of cognitive science is the opposition between symbolic representation and situated activity as accounts of the human mind. Norman (1993) provides a simplified characterization of the two positions: those who would model thought by emphasizing the computational capacities of the brain and those who would account for thought in the "mutual accommodation of people and the environment" (1993:4). Supporters of the former position are said to discount external factors in developing theories of the mind – social, cultural, historical, and environmental influences that proponents of the latter position argue are the primary structures of thought.

An anthropology of knowledge finds itself indebted to scholars of both traditions. We envision human agents with the cognitive capacities and predispositions that constitute the common architecture of the mind. Using these abilities, actors construct mental models from experience. These models in turn facilitate and are potentially modified in future activities. The human cognitive architecture simultaneously constrains and provides the potential for knowledge representation. Yet specific representational structures are derived from and applied in the everyday contexts of human behavior.

One aspect of the debate as we understand it may revolve around a distinction not commonly made between modular capacities and domain-specific knowledge (although see Hirschfeld and Gelman 1994). Humans are innately endowed with the capacities to process (among other stimuli) visual, aural, verbal, and kinesthetic information. Normal exposure to these respective data leads developing humans to construct appropriate information-processing mechanisms. These modular capacities become established early in life and persist throughout adulthood in relatively unchanged form.

Domain-specific theories, however, may be constructed and revised throughout life. These are the folk theories humans construct to explain experience in the world. These knowledge structures consist of governing principles and an encyclopedic range of schemata relevant to an empirical domain. The constructed folk theories are multimodal, drawing on the modular capacities as appropriate for representing information relevant to the experiences at issue. Domain-specific theories are substantively integrated conceptual systems oriented to interpretation and accomplishment in the real world. Social, cultural, historical, environmental, and material events provide the substance for building domain-specific theories. Both symbolic representation and situated action are essential to an adequate account of the acquisition and application of domain-specific knowledge. The work of Engeström (1987, 1993), Holland and Eisenhart (1990), Hutchins (1980), Murphy and Medin (1985), Quinn (1985), and others, taken in conjunction with this volume, provides evidence of the potential gains that may arise from integrating the contributions of these opposing perspectives.

More specific to anthropology is the opposition of perspectives on culture as public ideology and culture as knowledge. This debate takes shape in the criticism leveled by Clifford Geertz against a cognitive anthropology inspired by Ward Goodenough, who argued that culture is located in the minds and hearts of men (1981:51). In contrast, Geertz points out that "culture is public because meaning is" (1973:12). Geertz goes on to insist that "culture is not a power, something to which social events, behaviors, institutions or processes can be causally attributed; it is a context, something within which they can be intelligibly – that is thickly – described" (1973:14).

Yet the work of numerous anthropologists belies the exclusivity of cognitivist and interpretivist positions (see Sperber 1974; Bloch 1985; Boyer 1993; Shore 1991; Dougherty and Fernandez 1981, 1982). Scholars such as those just mentioned have begun to synthesize these opposing perspectives in a dynamic view of culture that emphasizes the process of learning in public contexts through private reflections and sedimentations and the process of anticipating and governing behavior on the basis of prior knowledge. In these perspectives knowledge is indeed private yet constructed from public contexts. Cultural ideologies are public structures, but dependent for their reproduction or alteration on the practices of individuals acting on their constructed systems of knowledge. The "imaginative universe" (Geertz 1973:13) exists in the practices of peoples and in the knowledge they have acquired through those practices. Perhaps we must resign ourselves to the duality of culture, in minds and in performances, if we are to capture its complexity with accuracy.

The anthropology of knowledge that we envision encourages us to transcend these debates in our own work. Each side of the oppositions identified above contributes crucially to a holistic framework. It is not a grievous error to see mind and practice simultaneously responsible for cultural phenomena if one recognizes that this requires dealing with different methods of description and analysis. We aim in our own work to be neither blatantly social nor fundamentally psychological: we have taken as our central problem the explication of the interdependence of principled cultural stability and emergent individual creativity (Leroi-Gourhan 1993).

Implications of an anthropology of knowledge

In these concluding pages we intend to comment on the implications of our research. We touch on concerns such as the uniquely human proclivity for tool use, *tool* here meaning the computer as well as the hammer; the embeddedness of skill in conceptual reasoning; mutability, constraint, and their interrelationship; the nature of the learning process; and the diversity of ways in which knowledge is represented.

Tool use

The first implication of the anthropology of knowledge developed here is the foundation it establishes for investigating cognition in the context of naturally occurring events (Engeström 1993; Hutchins 1980, 1993a, 1995; Holland and Quinn 1987; Holland and Eisenhart 1990; Lave 1988; Lave and Wenger 1991; Suchman 1987; Wood, in progress).

Instrument manipulation in the broadest sense is integral to human goal attainment. Both sophisticated knowledge structures and skillful behavior are entailed in the tool-mediated productions of everyday activities. It is essential to examine events in the real world to gain access to the complex nature of both the mental and enacted processes involved.

As was anticipated by Brown, Collins, and Duguid (1989:33),

> People who use tools actively . . . build an increasingly rich implicit understanding of the world in which they use the tools and of the tools themselves. The understanding, both of the world and of the tool, continually changes as a result of their interaction. Learning and acting are interestingly indistinct, learning being a continuous, life long process resulting from acting in situations.

The study of the real-world tool-mediated integration of knowledge and practice provides evidence of multimodal conceptualization and the emergent character of human thought and activity. Controlled research contexts typically provide the advantage of more precise specification of the variables relevant to a subject under investigation or a hypothesis being tested. Yet it is perhaps just this factor of control that may preclude the potential to grasp the multifaceted human use of mind and body in naturally occurring situations. A focus on naturally occurring tool use and accomplishment opens up the range of questions for which a cognitive science enterprise can begin to provide interesting answers.

Mind and body in skilled performance

Another implication of the anthropology of knowledge we develop here is the embedding of skill in conceptual reason. In the company of scholars like Douglas Harper (1987), we have argued for a unity of work, a marriage of the hand and mind as an essential dimension of human activity. The mind–body dualism typical of Western science and philosophy has failed to account for the subjective experience of unity in productive activity. The separation of mind and body in research and the dissolution of activity systems in scholarly investigation also fails to provide accounts of the human realization of fulfillment in accomplishment.

An anthropology of knowledge provides a framework for understanding skilled performances as conceptually embedded even when immediate events press an agent to react seemingly without thinking. Even the most urgent embodied responses are presaged by verbal, visual, or kinesthetic representations derived from prior activities. Clearly, issues of cognitive processing remain to be clarified here. But behavioristic notions of unreflected physical talents or bodily intuitions are inadequate as full accounts of physical actions and reactions in contexts of skilled performance.

Mutability and constraint

A third implication of our framework for research is the potential in it to address the interplay of mutability and constraint characteristic of human behavior. The specification of relations among structural principles and real behaviors becomes possible in accounting for continuity and change as properties of both synchronic and historical events.

Productive activity is creative, innovative, and flexible. Its explanation lies in something more like Sahlins's structure of the conjuncture (1985) than in either the notion of replication of templates or culture grammar. Ideas acquired through prior experiences are at risk as subsequent experiences unfold and practices are at risk in reflected thought. The process is one of constant negotiation of enactment and understanding wherein structural resources and the results of prior actions inform activity, which, in turn, is a resource for revising those very structures. As a result, a human acting in the world is scientist and *bricoleur* in Lévi-Strauss's terms: operating as "scientist creating events (changing the world) by means of structures" and as " 'bricoleur' creating structures by means of events" (1966:22). This is simultaneously the crux of emergence and the context in which knowledge and practice can be construed as both principled and protean.

This account belies explanations of craftsmanship as reproduction of previous events relying on authoritative knowledge, mental templates, and rules as the structuring elements. It belies the *chaîne opératoire* (Lemonnier 1992) and other procedural accounts of behavior that lack a governing conceptual dimension and requires that we account for technology and planful activity rather than simply describing techniques and acts (see also Ingold 1994).

An anthropology of knowledge transcends even the generative potential of the cultural analogy to linguistic grammars. Humans as producers of artifacts other than speech do not seem to establish formal and ultimately unchanging grammars for organizing their behavior, as they do for their language use. Instead they construct domains governed by principles and stocks of schemata generalized from events. While increasingly stable as an individual becomes increasingly experienced in an activity, such knowledge systems allow greater latitude in the formal arrangement of elements and are always subject to revisions rooted in reflections that occur throughout life.

In this way an individual enters the larger processes of society and history. Productions do not feed predictably back into the principles and schemata from which they emerge but contain the potential to reaffirm or restructure those ideas. Private reaffirmations and changes in structure become open to public consideration as these ideas are expressed in the practices and products of human activity.

Sperber suggests that knowledge might be studied through an "epidemiology of representations" (1985) that would explore the incidence and distribution of information within and across communities. This is a tantalizing challenge that would go beyond the traditional conundrums posed by issues of consensus and variation for theories of culture. The challenge to provide an epidemiology of ideas should force us to examine the changing relations among thoughts, practices, and products as distributed among practitioners over time and space. Ultimately such research should contribute significantly to dynamic and processual notions of culture.

Learning

The anthropology of knowledge developed herein has implications as well for a theory of learning founded simultaneously in cognitive predispositions and situated action (Hirschfeld and Gelman 1994; Hutchins 1980, 1993a; Lave and Wenger 1991). The cognitive architecture of the human mind predisposes an actor to understand experience in terms of entities and their relations (Hutchins 1980). Universally available conceptual distinctions probably underlie the construction of particular folk theories of experience (Hirschfeld and Gelman 1994). This is not to "pretend a science that cannot be found" (Geertz 1973:20), but to acknowledge a foundation (not an end product) for learning.

Building on the cognitive architecture to construct specific systems of knowledge is coincident with the development of efficacious performance. The dialectic between the mental and physical dimensions of activity creates problematic contexts that result in learning. The situations in which human agents perform are goal-oriented and learning can be measured, at least in part, by situated accomplishment.

One additional consequence of the situated nature of learning derives from the emergent property of activity. "Brilliant unforeseen results" (Lévi-Strauss 1966:17) are the potential outcome of each and every productive event, with the effect that learning from doing is possible throughout the life course. Once constructed by an actor, a set of principles governing an activity system is like a scientific paradigm in that the system is resistant to change. A person acting in the world will attempt to understand events in terms of this preestablished knowledge. However, the material results of actions, potentially unforeseen and unpredictable, may trigger reflections that lead to reorganizations of and novel introductions to the knowledge base.

The learning described here is a process of socialization that never ends (Lave and Wenger 1991). It is lifelong, open-ended, goal-oriented, and richly dependent upon experience. The problem-oriented nature of activity guides the learner-practitioner. Ultimately the learner may come to assume fully the identity of a practitioner; yet, in the case of artist-blacksmithing and perhaps other domains, this identity itself includes a value on open structures and novel events that insures continued learning in practice.

We recognize two cognitive strategies that characterize situated learning. One is the process of generalizing schemata from sets of similar experiences. This is one process of sedimentation to which Schutz refers. Empirical generalizations as schemata form the encyclopedic substance of a domain. In the case of artist-blacksmithing, this is the stock of information held in mind about the shop, tools, procedures, artifacts, aesthetics, mechanics, and social others.

The second process by which knowledge becomes sedimented is that of the abstraction of defining principles from abundant experience. This is the process of defining an activity itself, of constructing a folk theory of a domain, in terms of a set of principles that guide the formation of goal orientations, evaluations, plans, and production. This process is slower than the building of schemata, often requiring long-term experience in a domain (Gatewood 1985). The incremental acquisition of diverse schemata for production and comparative consideration of a growing encyclopedia of knowledge results in the eventual sedimentation of principles; these principles are subsequently drawn upon to govern activity that may yield further substantive schemata. This sedimentary process often proceeds out of the awareness of the human actor and comes to take the form of taken-for-granted premises for organizing work. Yet the governing principles thereby derived are within the reach of understanding and reflection and constitute the foundation for coherent comprehension of an activity system.

This perspective on learning allows us to address difficulties in the transfer of acquired knowledge often noted by learning theorists. Once a basic theory of a domain is constructed, principles encompass the encyclopedia of schemata. New knowledge is acquired in context and sedimented in relation to prior knowledge of one domain or another. As a consequence, knowledge is typically called up in situations constituted by the activities from which it derived.

Transfer of schemata or principles of a given domain to another will be constrained by the relative exclusivity of the respective empirical contexts for learning and acting. The knowledge acquired for forging iron would not necessarily transfer easily to the activities of writing or swimming or getting married. Such transfers or generalizations across domains may occur, but require constructing partial analogies between activity systems that ignore empirical and principled differences and focus on similarities in abstract relations or imagery. One such generalization might be the concern for how to end a line, equally apparent among artist-blacksmiths, musical composers, and performers of vocal and instrumental music.

Lave has referred to situations in which knowledge and practice converge as conferring ownership on practitioners – ownership in the sense of a holistic conceptual and practical competence (1988:156). The coherence of such a system constrains the simple transfer of isolated bits of information.

Diverse modalities of knowledge representation

Finally, attention to the diversity of cognitive modes through which knowledge is represented opens research to questions of the integrated and specialized capacities of the mind. We have suggested that conceptualization is a product of diverse forms of information – diverse in their level of generality (principles versus schemata) and diverse in modes of representation (including linguistic, aural, visual, and kinesthetic systems) (see Jackendoff 1989; Inoue 1993). An actor's development of a coherent conceptual structure requires an ability to construct and move among diverse informational structures. Such movement is guided by correspondences in the distinct informational resources and by the ability to translate some information from one representational mode into another.

However, conceptualization retains the complexity of a multimodal system in which the diverse organizations of knowledge play distinct functional roles. In the case of constructed domain-specific knowledge, imaged representations provide both visual and kinesthetic information used in reasoning about future and ongoing activities and in governing practice. Detailed visual depictions and envisioned bodily actions provide a foundation for reasoning about objects and their components, spatial relations, and potential transformations. Such reasoning follows the logics (mechanical, physical, aesthetic) required by a task at hand and is necessary for planful activity. Propositional capacities, by contrast, facilitate categorization and schema construction, providing standard concepts for reference and communication. Principles that define an activity system and govern practice are formal abstractions. Expressions of the principles can be found in verbal, visual, kinesthetic, and material forms. An anthropology of knowledge directs us to investigate the diverse possibilities for representing and expressing information and to clarify the potential for integration of this information in conceptual thought and practice.

Bibliography

Anderson, James

1993 Charcoal Bluing of Rifle Barrels. *Journal of Historical Armsmaking Technology* 5:53–64.

Andreae, Christopher

1993 Calligraphy in Steel. *The Home Forum: Christian Science Monitor,* November 8.

Andrews, Jack

1977 *Edge of the Anvil: A Resource Book for the Blacksmith.* Emmaus, PA: Rodale Press.

1992 *Samuel Yellin, Metalworker.* Ocean City, MD: Skipjack Press.

Appleton, D.

1852 *Appleton's Dictionary of Machines, Mechanics, Engine-work and Engineering.* New York: D. Appleton.

Arnold, Dean

1985 *Ceramic Theory and Cultural Process.* Cambridge University Press.

Bacon, John Lord

1986 [1904, 1908] *Elementary Forge Practice.* Reprinted in Lost Technology Series. Bradley, IL: Lindsay Publications.

Barker, Roger G.

1968 *Ecological Psychology: Concepts and Methods for Studying the Environment of Human Behavior.* Stanford, CA: Stanford University Press.

Bateson, Gregory

1958 *Naven.* Second edition. Stanford, CA: Stanford University Press.

Bealer, Alex

1976 *The Art of Blacksmithing.* Revised edition. New York: Funk and Wagnalls.

The Blacksmith and Wheelwright

1910 Advertisement for Mayer Bros. Company Manufacturers, Mankato, MN. December issue. New York: Richardson Co.

Bloch, Maurice

1985 From Cognition to Ideology. In Richard Fardon (ed.), *Knowledge and Power: Anthropological and Sociological Approaches.* Edinburgh: Scottish Academic Press.

Bourdieu, Pierre
1977 *Outline of a Theory of Practice.* Cambridge University Press.
Boyer, Pascal (ed.)
1993 *The Cognitive Implications of Religious Symbolism.* Cambridge University Press.
Brown, John Seely, Allan Collins, and Paul Duguid
1989 Situated Cognition and the Culture of Learning. *Educational Researcher,* January–February, 32–41.
Brumfield, Gary
1993 Cock Forging: A Study in Technology. *Journal of Historic Armsmaking Technology* 5:65–87.
Bunzel, Ruth
1929 *The Pueblo Potter.* New York: Columbia University Press. Reprinted by Dover, 1972.
Carey, Susan, and Elizabeth Spelke
1994 Domain-Specific Knowledge and Conceptual Change. In Lawrence A. Hirschfeld and Susan A. Gelman (eds.), *Mapping the Mind,* pp. 169–201. Cambridge University Press.
Carlson, W. Bernard
1991 Building Thomas Edison's Laboratory at West Orange, New Jersey. *History of Technology* 13:150–167.
Centaur Forge Ltd.
1993 Catalog. May. Burlington, WI.
Chaiklin, Seth, and Jean Lave (eds.)
1993 *Understanding Practice: Perspectives on Activity and Context.* Cambridge University Press.
Cochran, Thomas C.
1981 *Frontiers of Change.* New York: Oxford University Press.
Coles, John
1973 *Archaeology by Experiment.* New York: Charles Scribner's Sons.
Cooper, Carolyn C.
1991a Visualizing Invention from Tools to Machines in the Career of Thomas Blanchard (1788–1864). *Chronicle of the Early American Industries Association* 44(1):3–10.
1991b *Shaping Invention: Thomas Blanchard's Machinery and Patent Management in Nineteenth-Century America.* New York: Columbia University Press.
Cooper, Eugene
1989 Apprenticeship as a Field Method: Lessons from Hong Kong. In Michael W. Coy (ed.), *Apprenticeship: From Theory to Method and Back Again.* Albany: State University of New York Press.
Coy, Michael W. (ed.)
1989 *Apprenticeship: From Theory to Method and Back Again.* Albany: State University of New York Press.

D'Andrade, Roy
1989 Cultural Cognition. In M. Posner (ed.), *Foundations of Cognitive Science,* pp. 795–830. Cambridge, MA: MIT Press.
1991 Connectionism and Culture, or Some Implications of a New Model of Cognition with Respect to Anthropological Theory. Paper delivered at the meeting for the Society for Psychological Anthropology.
1992 Schemes and Motivation. In Roy D'Andrade and Claudia Strauss (eds.), *Human Motives and Cultural Models,* pp. 23–45. Cambridge University Press.

D'Andrade, Roy, and Claudia Strauss (eds.)
1992 *Human Motives and Cultural Models.* Cambridge University Press.

Deetz, James
1967 *Invitation to Archaeology.* Garden City, NY: Natural History Press.

Dougherty, J. W. D. (ed.)
1985 *Directions in Cognitive Anthropology.* Urbana-Champaign: University of Illinois Press.

Dougherty, J. W. D., and James Fernandez
1981 Introduction. In Clark E. Cunningham, Janet W. D. Dougherty, James W. Fernandez, and Norman E. Whitten (eds.), Special Issue on Symbolism and Cognition I. *American Ethnologist* 8(3):413–421.
1982 Afterword. In Janet W. D. Dougherty, James W. Fernandez, Emiko Ohnuki-Tierney, and Norman E. Whitten (eds.), Special Issue on Symbolism and Cognition II. *American Ethnologist* 9(4):820–832.

Dougherty, J. W. D., and Charles Keller
1982 Taskonomy: A Practical Approach to Knowledge Structures. Special Issue on Symbolism and Cognition II. *American Ethnologist* 9(4):763–774.

Drew, J. M.
1915 *Farm Blacksmithing.* St. Paul, MN: Webb.

Engeström, Yrgö
1987 *Learning By Expanding: An Activity-Theoretical Approach to Developmental Research.* Helsinki: Orienta-Konsultit Oy.
1993 Developmental Studies of Work as a Testbench of Activity Theory: The Case of Primary Care Medical Practice. In Seth Chaiklin and Jean Lave (eds.), *Understanding Practice: Perspectives on Activity and Context.* Cambridge University Press.

Ewbank, Thomas
1849 *A Descriptive and Historical Account of Hydraulic and other Machines.* New York: Greeley and McElrath.

Ferguson, Eugene S.
1973 Technology as Knowledge. In Edwin T. Layton, Jr. (ed.), *Technology as Social Change.* New York: Harper and Row.
1977 The Mind's Eye: Nonverbal Thought in Technology. *Science* 197:827–836.

Frake, Charles
1985 Cognitive Maps of Time and Tide Among Medieval Seafarers. *Man* n.s. 20:254–270.
Friese, John F.
1921 *Farm Blacksmithing.* Peoria, IL: Manual Arts Press.
Gatewood, John
1985 Actions Speak Louder Than Words. In J. W. D. Dougherty (ed.), *Directions in Cognitive Anthropology,* pp. 199–220. Urbana-Champaign: University of Illinois Press.
Geertz, Clifford
1973 *The Interpretation of Cultures.* New York: Basic Books.
Gerakaris, D.
1993 At the Founding: The Birth of ABANA. *The Anvil's Ring* 21(2):10.
Gibson, J. J.
1977 The Theory of Affordances. In R. E. Shaw and J. Bransford (eds.), *Perceiving, Acting and Knowing.* Hillsdale, NJ: Erlbaum.
Giddens, Anthony
1976 *New Rules of the Sociological Method.* New York: Basic Books.
Gladwin, Thomas
1970 *East Is a Big Bird: Navigation and Logic on Puluwat Atoll.* Cambridge, MA: Harvard University Press.
Goffman, Erving
1974 *Frame Analysis.* New York: Harper and Row.
Goodenough, Ward
1957 Cultural Anthropology and Linguistics. In Paul L. Garvin (ed.), Report of the Seventh Annual Round Table Meeting on Linguistics and Language Study. Monograph Series on Languages and Linguistics, no. 9, pp. 167–173. Washington, DC: Georgetown University Press.
1981 *Culture, Language and Society.* Second edition. Menlo Park, CA: Benjamin Cummings.
Goodwin, Charles, and Marjorie Harness Goodwin
1992 Context, Activity and Participation. In Peter Auer and Aldo di Luzio (eds.), *The Contextualization of Language,* pp. 77–100. Philadelphia: John Benjamins.
Googerty, Thomas F.
1911 *Hand-Forging and Wrought-Iron Ornamental Work.* Chicago: Popular Mechanics Co.
Gopnick, Alison, and Henry M. Wellman
1994 The Theory Theory. In Lawrence A. Hirschfeld and Susan A. Gelman (eds.), *Mapping the Mind,* pp. 294–316. Cambridge University Press.
Gordon, Robert B., and Patrick Malone
1994 The Texture of Industry: An Archaeological View of the Industrialization of North America. New York: Oxford University Press.

Gorman, Michael E., and W. Bernard Carlson

1990 Interpreting Invention as a Cognitive Process: The Case of Alexander Graham Bell, Thomas Edison, and the Telephone. *Science, Technology and Human Values* 15(2):131–164.

1992 Mapping, Invention and Design: The Invention of the Telephone Gives Insight on the Process of Technological Creativity. *Chemtech,* October:584–591.

Greeno, James G. (ed.)

1993 *Situated Action.* Special Issue of *Cognitive Science* 17(1):1–147.

Greeno, James G., and Joyce Moore

1993 Situativity and Symbols: Response to Vera and Simon. In James G. Greeno (ed.), *Situated Action.* Special Issue of *Cognitive Science* 17(1):49–60.

Habermas, Jurgen

1971 *Knowledge and Human Interests.* Boston: Beacon Press.

Hanks, William

1991 Foreword. In Jean Lave and Etienne Wenger, *Situated Learning: Legitimate Peripheral Participation.* Cambridge University Press.

Hansen, Carl L., and Charles M. Keller

1971 Activity Patterning and Context, Isimila Korongo, Iringa district, Tanzania. *American Anthropologist* 73(5):1201–1210.

Hardin, Margaret

1977 Individual Style in San José Pottery Painting: The Role of Deliberate Choice. In James N. Hill and Joel Funn (eds.), *The Individual in Prehistory.* New York: Academic Press.

1984 Models of Decoration. In Sander E. van der Leeuw and Alison Pritchard (eds.), *The Many Dimensions of Pottery: Ceramics in Archaeology and Anthropology. Cingula* 7:575–607. Amsterdam: University van Amsterdam Albert Egges van Giffen Institut voor Prae- en Protohistory.

Harper, Douglas

1987 *Working Knowledge: Skill and Community in a Small Shop.* Chicago: University of Chicago Press.

Hasluck, Paul (ed.)

1912 *Smiths' Work.* New York: Cassell.

Hill, Jacquetta, and David Plath

1996 Moneyed Knowledge and Becoming a Master Diver. In John Singleton (ed.), *Learning in Likely Places.* Cambridge University Press.

Hill, James N.

1994 Prehistoric Cognition and the Science of Archaeology. In Colin Renfrew and Ezra B. W. Zubrow (eds.), *The Ancient Mind: Elements of Cognitive Archaeology.* Cambridge University Press

Hindle, Brooke

1981 *Emulation and Invention.* New York: New York University Press.

Hindle, Brooke, and Steven Lubar
1986 *Engines of Change: The American Industrial Revolution 1790–1800.* Washington, DC: Smithsonian Institution Press.
Hirschfeld, Lawrence A., and Susan A. Gelman (eds.)
1994 *Mapping the Mind: Domain Specificity in Cognition and Culture.* Cambridge University Press.
Holland, Dorothy
1992 The Woman Who Climbed Up the House: Some Limitations of Schema Theory. In Theodore Schwartz, Geoffrey M. White, and Catherine A. Lutz (eds.), *New Directions in Psychological Anthropology,* pp. 68–82. Cambridge University Press.
Holland, Dorothy, and Margaret A. Eisenhart
1990 *Educated in Romance: Women, Achievement and College Culture.* Chicago: University of Chicago Press.
Holland, Dorothy, and Naomi Quinn (eds.)
1987 *Cultural Models in Language and Thought.* Cambridge University Press.
Holmstrom, J. G.
1900 *Modern Blacksmithing.* Chicago: Alhambra Book Co. Reprinted by Lindsey Publications.
Hunn, Eugene
1985 The Utilitarian Factor in Folk Biological Classification. In J. W. D. Dougherty (ed.), *Directions in Cognitive Anthropology,* pp. 141–160. Urbana-Champaign: University of Illinois Press.
Hutchins, Edwin
1980 *Culture and Inference: A Trobriand Case Study.* Cambridge, MA: Harvard University Press.
1983 Understanding Micronesian Navigation. In Dedre Gentner and Albert L. Stevens (eds.), *Mental Models,* pp. 191–225. Hillsdale, NJ: Erlbaum.
1993a Learning to Navigate. In Seth Chaiklin and Jean Lave (eds.), *Understanding Practice: Perspectives on Activity and Context,* pp. 35–63. Cambridge University Press.
1993b Review of Jean Lave and Etienne Wenger, *Situated Learning: Legitimate Peripheral Participation. American Anthropologist* 95(3):743–744.
1995 *Cognition in the Wild.* Cambridge, MA: MIT Press.
Ingersoll, D., J. E. Yellen, and W. MacDonald
1977 *Experimental Archaeology.* New York: Columbia University Press.
Ingold, Tim
1994 Tool Using, Tool Making and the Evolution of Language. In Duane Quiatt and Junichiro Itani (eds.), *Hominid Culture in Primate Perspective,* pp. 279–314. Niwot: University Press of Colorado.
Inoue, Kyoko
1993 Japanese Numeral Classifiers: Their Implications for Conceptual Coherence. *Belgian Journal of Linguistics* 8:57–77.

Jackendoff, Ray
1989 *Consciousness and the Computational Mind.* Cambridge, MA: MIT Press.
1992 *Languages of the Mind.* Cambridge, MA: MIT Press.

Johnson, Mark
1987 *The Body in the Mind.* Chicago: University of Chicago Press.

Jordan, Brigitte
1993 *Birth in Four Cultures.* Fourth edition. Revised and expanded by Robbie Davis-Floyd. Prospect Heights, IL: Waveland Press.

Karlin, C., and M. Julien
1994 Prehistoric Technology: A Cognitive Science. In C. Renfrew and E. B. W. Zubrow (eds.), *The Ancient Mind: Elements of a Cognitive Archaeology,* pp. 152–164. Cambridge University Press.

Keesing, Roger
1979 Linguistic and Cultural Knowledge. *American Anthropologist* 81(1):14–36.
1993 Earth and Path as Complex Categories: Semantics and Symbolism in Kwaio Culture. In Pascal Boyer (ed.), *Cognitive Aspects of Religious Symbolism.* Cambridge University Press.

Keller, Charles M.
1966 A Preliminary Report on the Development of Edge Damage Patterns on Stone Tools. *Man* 1(4):501–511.
1970 Montagu Cave: A Preliminary Report. *Quaternaria* 8:187–203.
1973 Montagu Cave in Prehistory: A Descriptive Analysis. *University of California Anthropological Records,* vol. 28.
1976 Notes from Apprenticeship to a Blacksmith, Santa Fe, New Mexico. Unpublished manuscript.
1978 Charles M. Keller. Faculty Interest Profiles. Department of Anthropology, University of Illinois, Urbana-Champaign.
1994 Invention, Thought and Process: Strategies in Iron Tool Production. In Sarah U. Wisseman and Wendell S. Williams (eds.), *Ancient Technologies and Archaeological Materials,* pp. 59–70. Langhorne, PA: Gordon and Breach Science Publishers.

Keller, Charles M., Carl L. Hansen, and Charles S. Alexander
1975 An Investigation of the Archaeology and Paleoenvironments in the Manyara and Engaruka Basins, Northern Tanzania. *Geographical Review* 65(3):364–376.

Keller, Charles M., and Janet Dixon Keller
1991a Thinking and Acting with Iron. Beckman Institute for Advanced Science and Technology, Cognitive Science Technical Report CS-91-08.
1991b The Dynamics of Productive Activity. In *Proceedings of the Conference on Critical Problems and Research Frontiers in the History of Science and the History of Technology,* Madison, WI
1993 Thinking and Acting with Iron. In Seth Chaiklin and Jean Lave (eds.),

Understanding Practice: Perspectives on Activity and Context, pp. 125–143. Cambridge University Press.

1996 Imaging in Iron or Thought Is Not Inner Speech. In John Gumperz and Stephen Levinson (eds.), *Rethinking Relativity. Proceedings of the 112th International Wenner-Gren Conference on Rethinking Linguistic Relativity, Jamaica, 1991.* Cambridge University Press.

Keller, Janet Dixon

1988 Woven World. *Mankind* 18(1):1–13.

Keller, Janet Dixon, and F. K. Lehman

1991 Complex Concepts. *Cognitive Science* 15(2):271–292.

Kline, Stephen J.

1984 Innovation is Not a Linear Process. *Research Management* 28:36–45.

Knight, Edward

1877 *Knight's American Mechanical Dictionary.* Cambridge, MA: Hurd and Houghton University Press.

Koch, Ursula

1977 Das Reihengraberfeld Bei Schretzheim. *Germanische Denkmaler Der Volkwanderungszeit;* Series A, BD, 13. Berlin: Gebr. Mann Verlag.

John Michael Kohler Arts Center

1980 *The Metalwork of Albert Paley.* Sheboygan, *WI.*

Kolers, Paul A., and William E. Smyth

1979 Images, Symbols and Skills. *Canadian Journal of Psychology* 33(3):158–184.

Kosslyn, S. M.

1980 *Image and Mind.* Cambridge, MA: Harvard University Press.

1981 The Medium and the Message in Mental Imagery: A Theory. *Psychological Review* 88(1):46–66.

1987 Seeing and Imaging in the Cerebral Hemispheres: A Computational Approach. *Psychological Review* 94(2):148–175.

Krause, Richard

1985 *The Clay Sleeps.* University: University of Alabama Press.

Lakoff, George

1987 *Women, Fire and Dangerous Things.* Chicago: University of Chicago Press.

Lasansky, J.

1980 *To Draw, Upset and Weld: The Work of the Pennsylvania Rural Blacksmith 1742–1935.* Lewisburg, PA: Oral Traditions Project of the Union County Historical Society.

Latané, Tom

1993 The Making of a Chest. *The Anvil's Ring* 21(3):22.

Lave, Jean

1988 *Cognition in Practice: Mind, Mathematics and Culture in Everyday Life.* Cambridge University Press.

Lave, Jean, and Etienne Wenger

1991 *Situated Learning: Legitimate Peripheral Participation.* Cambridge University Press.

Lechtman, Heather

1984 Andean Value Systems and the Development of Prehistoric Metallurgy. *Technology and Culture* 25:1–36.

Lemonnier, Pierre

1992 *Elements for an Anthropology of Technology.* Anthropological Papers no. 88. Museum of Anthropology, University of Michigan, Ann Arbor, MI.

Lemonnier, Pierre (ed.)

1993 Technological Choices: Transformations in Material Cultures Since the Neolithic. New York: Routledge.

Leont'ev, A. N.

1981 The Problem of Activity in Psychology. In J. V. Wertsch (ed. and trans.), *The Concept of Activity in Soviet Psychology,* pp. 37–71. Armonk, NY: M. E. Sharpe. Reprinted from *Voprosy Filosofi* (1972) 9:95–108.

Leroi-Gourhan, André

1993 *Gesture and Speech.* Translated from the French by Anna Bostock Berger. Cambridge, MA: MIT Press.

Leslie, Alan M.

1994 ToMM, ToBY and Agency: Core Architecture and Domain Specificity. In Lawrence A. Hirschfeld and Susan A. Gelman (eds.), *Mapping the Mind: Domain Specificity in Cognition and Culture,* pp. 119–148. Cambridge University Press.

Lévi-Strauss, Claude

1966 *The Savage Mind.* Chicago: University of Chicago Press.

Link, Carol A. B.

1975 Japanese Cabinetmaking: A Dynamic System of Decisions and Interactions in a Technical Context. Ph.D. Dissertation, Department of Anthropology, University of Illinois, Urbana-Champaign. Ann Arbor, MI: University Microfilms.

Lubar, Steven

1987 Culture and Technological Design in the 19th Century Pin Industry: John Howe and the Howe Manufacturing Company. *Technology and Culture* 28(2):253–282.

Lutz, Catherine

1992 Motivated Models. In Roy D'Andrade and Claudia Strauss (eds.), *Human Motives and Cultural Models,* pp. 181–190. Cambridge University Press.

Marr, David

1982 *Vision.* San Francisco: Freeman.

Maryon, H.

1936 *Soldering and Welding in the Bronze and Early Iron Ages.* Technical Studies in the Field of the Fine Arts. Fogg Art Museum, Harvard University.

1938 The Technical Methods of the Irish Smiths in the Bronze and Early Iron Ages. *Proceedings of the Royal Irish Academy, Section C* 44:7.

1941 Archaeology and Metallurgy I, Welding and Soldering. *Man* 41:118–126.

McNaughton, Patrick R.
1988 *The Mande Blacksmiths: Knowledge, Power and Art in West Africa.* Bloomington: Indiana University Press.

Medin, Doug
1989 Concepts and Conceptual Structure. *American Psychologist* 44(12):1469–1481.

Meilach, Dona Z.
1977 *Decorative and Sculptural Ironwork.* New York: Crown Publishers.

Miller, George A., Eugene Galanter, and Karl H. Pribram
1960 *Plans and the Structure of Behavior.* New York: Holt, Rinehart and Winston.

Moxon, Joseph
1975 [originally published 1703] *Mechanick Exercises.* Scarsdale, NY: Early American Industries Association.

Murphy, Gregory L., and Douglas L. Medin
1985 The Role of Theories in Conceptual Coherence. *Psychological Review* 92(3):289–316.

Narayan, Kirin
1993 How Native Is a "Native" Anthropologist? *American Anthropologist* 95(3):671–686.

Nolan, Charles
n.d. *Yesterday's Blacksmith.* Privately published.

Norman, Donald
1988 *The Design of Everyday Things.* New York: Doubleday Currency.
1993 Cognition in the Head and in the World: An Introduction. In James G. Greeno (ed.), *Situated Action.* Special Issue of *Cognitive Science* 17(1):1–6.

O'Neale, Lila M.
1932 Yurok-Karok Basket Weavers. *University of California Publications in American Archaeology and Ethnology* 32(1):1–184.

Ortner, Sherry B.
1989 *High Religion: A Cultural and Political History of Sherpa Buddhism.* Princeton: Princeton University Press.

Paley, Albert
1993 Paley's Tables. *The Anvil's Ring* 21(3):14.

Peacock, D. P. S.
1982 *Pottery in the Roman World: An Ethnoarchaeological Approach.* London: Longman.

Pfaffenberger, Bryan
1992 The Social Anthropology of Technology. *Annual Reviews in Anthropology* 21:491–516.

Pinker, Steven, and Paul Bloom
1990 Natural Language and Natural Selection. *Behavioral and Brain Sciences* 13:707–784.

Pirsig, R.
1974 *Zen and the Art of Motorcycle Maintenance: An Inquiry into Values.* New York: William Morrow.

Polanyi, Michael
1962 *Personal Knowledge.* Chicago: University of Chicago Press.

Pye, David
1968 *The Nature and Art of Workmanship.* Cambridge University Press.
1978 *The Nature and Aesthetics of Design.* London: Barrie and Jenkins.

Quinn, Naomi
1985 Commitment in American Marriage. In J. W. D. Dougherty (ed.), *Directions in Cognitive Anthropology,* pp. 291–320. Urbana-Champaign: University of Illinois Press.

Reichard, Gladys A.
1974 [originally published 1936] *Weaving a Navajo Blanket* (originally *Navajo Shepherd and Weaver).* New York: Dover.

Reichelt, Richard
1988 *Heartland Blacksmiths: Conversations at the Forge.* Carbondale: Southern Illinois University Press.

Renfrew, Colin
1994 Towards a Cognitive Archaeology. In Colin Renfrew and Ezra B. W. Zubrow (eds.), *The Ancient Mind: Elements of a Cognitive Archaeology.* Cambridge University Press.

Renfrew, Colin, and Ezra B. W. Zubrow (eds.)
1994 *The Ancient Mind: Elements of a Cognitive Archaeology.* Cambridge University Press.

Richardson, M. T.
1978 [originally published 1889–1891] *Practical Blacksmithing.* New York: Weathervane Books.

Rosch, E., C. B. Mervis, W. Gray, D. Johnson, and P. Boyes-Braem
1975 Basic Objects in Natural Categories. *Cognitive Psychology* 8:382–439.

Sahlins, Marshall
1985 *Islands of History.* Chicago: University of Chicago Press.

Sanders, Thomas
1993 Iron and Steel for the 1830's Indiana Blacksmith. *Midwest Open-Air Museums Magazine* 9:1.
1994 The Georgian Axe: A Study in Innovation. Paper presented to the Department of Anthropology, University of Minnesota, Minneapolis.

Schank, R., and R. Abelson
1977 *Scripts, Plans, Goals and Understanding.* Hillsdale, NJ: Erlbaum.

Schiffer, Michael Brian
1992 *Technological Perspectives on Behavioral Change.* Tucson: University of Arizona Press.

Schlanger, Nathan
1990 The Making of a Soufflé: Practical Knowledge and Social Senses. *Techniques et Culture* 15:29–52.
1994 Mindful Technology: Unleashing the Chaîne Opératoire for an Archaeology of Mind. In Colin Renfrew and E. B. W. Zubrow (eds.), *The Ancient Mind: Elements of a Cognitive Archaeology*, pp. 143–152. Cambridge University Press.
Schutz, Alfred
1967 *The Phenomenology of the Social World.* Translated by George Walsh and Frederick Lehnert. Evanston, IL: Northwestern University Press.
1971 *Collected Papers I. The Problem of Social Reality*. The Hague: Martinus Nijhoff.
Schwartzkopf, Ernst
1916 *Plain and Ornamental Forging*. New York: John Wiley and Sons.
Shanker, S. G.
1991 Muddles of Discovery. In M. Galbraith and W. J. Rapaport (eds.), *Where Does Insight Come From?* Technical Report 91-01, Center for Cognitive Science. Buffalo: State University of New York Press. Republished in 1995 as a Special Issue of *Minds and Machines* 5(4).
Shore, Bradd
1991 Twice Born, Once Conceived: Meaning Construction and Cultural Cognition. *American Anthropologist 93*(1):9–27.
Sieber, Roy
1972 *African Textiles and Decorative Arts*. New York: Museum of Modern Art.
Singleton, John (ed.)
1996 *Learning in Likely Places*. Cambridge University Press.
Smith, Cyril Stanley
1966 On the Nature of Iron. In *Made of Iron*. Exhibit Catalog. Houston: University of St. Thomas Art Department .
Southern Illinois University Museum and Art Galleries
1976 *Iron, Solid Wrought/USA*. Carbondale, IL.
Spears, Carol S.
1975 Hammers, Nuts and Jolts, Cobbles, Cobbles, Cobbles: Experiments in Cobble Technologies in Search of Correlates. In C. M. Baker (ed.), *The Arkansas Eastman Archaeological Project*. Arkansas Archaeological Survey, Research Report 6:83–116.
Sperber, Dan
1974 *Rethinking Symbolism*. Cambridge University Press.
1985 Anthropology and Psychology: Towards an Epidemiology of Representations. *Man* n.s. 20:73–89.
Sperber, Dan and Deidre Wilson
1986 *Relevance: Communication and Cognition*. Cambridge, MA: Harvard University Press.

Spradley, James and David McCurdy
1972 *The Cultural Experience: Ethnography in Complex Society*. Chicago: Science Research Associates.

Strauss, Claudia
1992 Models and Motives. In Roy D'Andrade and Claudia Strauss (eds.), *Human Motives and Cultural Models,* pp. 1–20. Cambridge University Press.

Suchman, Lucy A.
1987 *Plans and Situated Actions: The Problem of Human–Machine Communication.* Cambridge University Press.

Suchman, Lucy, and Randall H. Trigg
1993 Artificial Intelligence as Craftwork. In Seth Chaiklin and Jean Lave (eds.), *Understanding Practice: Perspectives on Activity and Context.* Cambridge University Press.

Trigger, Bruce
1991 Distinguished Lecture in Archaeology: Constraint and Freedom. *American Anthropologist* 93(3):551–569.

Uselding, Paul
1974 Elisha K. Root, Forging, and the "American System." *Technology and Culture* 15(4):543–569.

van der Leeuw, Sander E.
1989 Risk, Perception, Innovation. In Sander E. van der Leeuw and Robin Torrance (eds.), *What's New? A Closer Look at the Process of Innovation.* London: Unwin Hyman.
1993 Giving The Potter a Choice: Conceptual Aspects of Pottery Techniques. In Pierre Lemonnier (ed.), *Technological Choices: Transformations in Material Cultures Since the Neolithic,* pp. 238–289. New York: Routledge.

Van Esterik, P.
1985 Imitating Ban Chiang Pottery: Toward a Cognitive Theory of Replication. In J. W. D. Dougherty (ed.), *Directions in Cognitive Anthropology,* pp. 221–242. Urbana-Champaign: University of Illinois Press.

Vera, Alonso, and Herbert A. Simon
1993 Situated Action: A Symbolic Interpretation. In James G. Greeno (ed.), *Situated Action.* Special Issue of *Cognitive Science* 17(1):7–48.

Vincenti, Walter G.
1984 Technological Knowledge Without Science: The Innovation of Flush Riveting in American Airplanes, ca. 1930–ca. 1950. *Technology and Culture* 25:540–576.
1990 *What Engineers Know and How They Know It: Analytical Studies From Aeronautical History*. Baltimore: Johns Hopkins University Press.

Vlach, John Michael
1981 *Charleston Blacksmith: The Work of Philip Simmons.* Athens: University of Georgia Press.

Vygotsky, L. S.
1978 *Mind in Society: The Development of Higher Psychological Processes.* Edited by M. Cole, V. John-Steiner, S. Scribner, and E. Souberman. Cambridge, MA: Harvard University Press.
Wagner, Helmut (ed.)
1970 *Alfred Schutz on Phenomenology and Social Relations: Selected Writings.* Chicago: University of Chicago Press.
Wallace, A. F. C.
1978 *Rockdale.* New York: Alfred A. Knopf.
Webster's Third New International Dictionary
1981 Philip Babcock Gove (ed.). Springfield, MA: G. & C. Merriam Company.
Wertime, Theodore, and James Muhly
1980 *The Coming of the Age of Iron.* New Haven: Yale University Press.
Wertsch, J. V. (ed. and trans.)
1981 *The Concept of Activity in Soviet Psychology.* Armonk, NY: M. E. Sharpe.
Weygers, Alexander G.
1973 *The Making of Tools.* New York: Van Nostrand Reinhold.
1974 *The Modern Blacksmith.* New York: Van Nostrand Reinhold.
Whittaker, Francis
1986 *The Blacksmith's Cookbook: Recipes in Iron.* Vail, CO: Jim Fleming Publications.
Whitten, Norman, and Dorothea S. Whitten
1993 *Imagery and Creativity.* Tucson: University of Arizona Press.
Wigginton, Eliot
1979 Ironmaking and Blacksmithing. *Foxfire* 5. Garden City, NY: Anchor Books.
Wood, William
in progress To Learn Weaving Below the Rock: The Making of Weaving and Weavers in Teotitlan del Valle. Ph.D. Dissertation. University of Illinois at Urbana-Champaign.
Wylie, William N. T.
1990 *The Blacksmith in Upper Canada.* Ganaoque, Ontario: Longdale Press.
Wynn, Thomas
1989 *The Evolution of Spatial Competence.* Illinois Studies in Anthropology no. 17. Urbana-Champaign: University of Illinois Press.
1991 Tools, Grammar and the Archaeology of Cognition. *Cambridge Archaeological Journal* 1(2):191–206.
Zinchenko, V. P., and V. M. Gordon
1981 Methodological Problems in the Psychological Analysis of Activity. In J. V. Wertsch (ed. and trans.), *The Concept of Activity in Soviet Psychology.* Armonk, NY: M. E. Sharpe.
Zirngibl, Manfred
1983 *Seltene Afrikanische Kurzwaffen.* Grafenau Morsak Verlag.

Selected name index

Subject index